Belted Galloway Cattle Ranching

A Comprehensive Guide to Breeding, Nutrition and Management

Twila Weiss & JD Weiss

ISBN: 978-1-964391-03-8

Table of Contents

We dedicate this book in loving memory of JD's parents, Richard and Marilynn Weiss, for their unwavering love and support, and his grandparents, Harry and Frances Dorin, whose guidance shaped the person he is today.

Introduction

Welcome to Belted Galloway cattle ranching, a venture combining tradition, resilience, and a future of delicious, high-quality beef. Whether you're new to breeding, feeding, or managing a herd of these unique animals—this book has you covered. But let's take a moment to enjoy the Belted Galloway breed's unique beauty and history before we get into the details.

Belted Galloways, more often known as "Belties" and the "Oreo cows," are native to the Galloway area in the rough hills of southwestern Scotland (Belted Galloway Society, n.d.). Their origins may be traced to the 16th century, when they mastered the art of thriving in extreme environments (*Belted Galloway*, 2022; *Belted Galloway Society*, n.d.).

Belties are easily identifiable by their characteristic belted coat, which features a white belt in the middle and contrasts with their black, dun, or red coat. However, their allure extends deeper than meets the eye. These cattle are strong and sturdy, making them ideal for rough terrain. They don't require much shelter during the hot summers and cold winters thanks to their thick, double coat, which serves as a natural insulator (*Belted Galloway*, 2022).

In addition to their attractive appearance, Belties are well-known for their calm temperament. The fact that they are naturally polled (hornless) makes them safer for ranchers and animals alike, while handling. Ranching with these cattle is less stressful because of their quiet nature and strong maternal instinct.

Belted Galloways are unique in that they are very good hunters. Belties are great at utilizing a variety of grasses and roughage, unlike many other cattle breeds that need specific, high-maintenance diets. Ranchers looking for a sustainable and cost-effective method will find them outstanding because of how efficiently they transform this coarse feed into high-quality beef. Their capacity to consume a wide variety of plants is an important part of conservation grazing, which is why these animals are ideal for this function.

Now, let's shift our focus to the concept of sustainable ranching practices. These days, people care more and more about how their food choices affect the environment. When ranching in a sustainable way, the animals' happiness, the land's health, and the ranch's financial stability are all given first priority. Embracing these methods, Belted Galloway ranchers help create a food system that is more responsible and ethical.

Taking care of the livestock is an important part of being sustainable. Belties don't need as much feed as other breeds because of their inherent hardiness and aptitude for thriving on pastures. This lessens the need for environmentally damaging, intensive feed production methods. Their grazing patterns are so efficient that they reduce soil erosion and boost biodiversity by eating different kinds of plants.

A key component of sustainable ranching is the prudent management of land. Rotational grazing is possible with Belted Galloways because of their adaptability to a wide variety of plants. By using this method, pastures can be kept healthy and diverse in grasses and legumes, which helps to avoid overgrazing. Consequently, the cattle receive the best possible nourishment from these lush meadows, establishing a self-sustaining cycle.

Last but not least, ranching in a sustainable way helps the rancher weather economic storms. Feed and infrastructure expenses are reduced due to the low maintenance requirements of Belties. There is a chance to get into the premium market because of the rising demand for grass-fed beef. Ranchers of Belted Galloway cattle may secure their livelihoods for the future and help create a greener food chain by adopting sustainable techniques.

If you are interested in the fulfilling life of Belted Galloway ranching, this detailed guide will be an invaluable resource. We'll go over the nuts and bolts of the Beltie diet, discuss breeding techniques for a healthy and productive herd, and offer practical advice on managing the herd. You can start a ranching path that is financially rewarding and ecologically responsible by learning about the special traits of these extraordinary cattle and adopting sustainable techniques.

Chapter 1:

Introduction to Belted Galloway Cattle

(Natural-born inspectors: our cattle leave no stone unturned or untasted!)

In the world of cattle, where their coats come in all sorts of colors and designs, there's one breed that really catches the eye: the Belted Galloway. These captivating creatures really know how to make a statement with their distinctive black and white belts. They always manage to turn heads wherever they go. However, their appeal goes way beyond just looks.

Imagine this: a tough breed that comes from the rough terrain of Scotland, shaped by centuries of harsh conditions to become models of ruggedness and flexibility. These cattle have such thick double coats that they can easily handle any kind of weather, whether it's mild or extreme.

But there's actually more to the Belted Galloway than what you might think at first glance. There's a pretty interesting history, tradition, and utility behind their striking appearance. Starting off as cherished livestock in the misty highlands, these animals have come a long way to become a cherished breed all over the world. Their journey is a testament to the strong connection between humans and animals, built on a foundation of respect and admiration.

In this chapter, we'll take a closer look at where they come from, delve into what makes them special, and discover how they've made a lasting impact on the world of agriculture. Get ready to be enchanted by the beauty and personality of these unique animals as we learn about their history.

The Ancestral Roots and Development

The story of the Belted Galloway begins in the Galloway region of southwest Scotland. For a long time, the Galloway cattle freely wandered this area. They're an old breed that's famous for being tough and versatile. The cattle, mostly black but sometimes with red or dun variations, were a great fit for the challenging surroundings. Of course, they were naturally hornless, also known as polled, and this trait would be inherited by the Belted Galloway.

There's been a lot of back and forth about where the belted pattern actually came from. According to some historians, it seems like the white-belted animals were selectively bred over time, possibly emerging naturally within the Galloway population.

It is not quite clear where the Belted Galloway's distinctive pattern came from. Although the Galloway breed's belted trait gained popularity in Scotland in the 1600s [Belted Galloway, 2022], some speculate that it may have crossed with the Dutch Lakenvelder in the centuries that followed, when trade between the two countries was at its strongest. But there's no hard proof that this theory is correct. Scottish farmers took selective breeding to the next level, making sure to enhance the belted feature and solidify the breed's one-of-a-kind appearance. They really focused on the breed's unique coat and its ability to withstand tough conditions (*Belted Galloway*, 2022).

Geographic Origins and Global Spread

The Belted Galloway did really well in the harsh Scottish climate, which was just right for the not-so-great pastures and harsh winters of the region. The breed is known for its thick double coat, which provides great insulation. They have a low metabolic rate, allowing them to survive on very little food. These adaptations were really helpful for farmers in the Galloway region. It would have been tough for other breeds to handle the challenges.

In the 20th century, the Belted Galloway made its way beyond British shores. In the 1940s, the breed ended up crossing the Atlantic and settling down in North America (The Livestock Conservancy, n.d.). They became popular because they could thrive in different climates, from the chilly Canadian plains to the warmer parts of the United States. Belted Galloways can be seen in different countries across the globe, such as Australia, New Zealand, and Europe, which shows how popular they are worldwide.

Historical Significance in Agriculture

The Belted Galloway has been a key player in Scottish agriculture for many years. They were just what the farmers in the region needed: able to thrive with minimal resources in the challenging environment. They could easily munch on rough pastures and heather moorlands, where other breeds would have a hard time finding food. Being self-sufficient was really important for farmers who had to deal with tough winters and limited resources.

In addition to being easy to care for, Belted Galloways were also highly valued for their delicious meat. This particular breed is known for producing beef that is highly marbled, resulting in a delicious flavor and tender texture. In addition to their calm nature and how easy they are to handle, Scottish farms found them to be quite valuable.

The Belted Galloway has a rich history that goes beyond just their practical contributions. They became a representation of strength and flexibility, capturing the essence of the Scottish Highlands. Their distinct look solidified their position in Scottish culture, frequently portrayed in art and literature.

The story of the Belted Galloway is all about survival, adaptation, and enduring appeal. These cattle have come a long way, from their ancient Galloway ancestors to their current global presence. They have definitely made a name for themselves in the world of agriculture. Their history, tied to the beautiful Scottish countryside, shows how tough they are and how clever the farmers were in taking care of them. For those who appreciate both functionality and aesthetics, the Belted Galloway remains a breed that never fails to captivate. It has a way of winning over hearts and making a positive impact on the agricultural scene worldwide.

Characteristics and Traits

(Newborn only minutes old)

The Belted Galloway is easily recognizable with its eye-catching black and white coat. But behind their attractive appearance, there's a wide range of interesting qualities and attributes that have contributed to their long-standing popularity.

Physical Description

Belted Galloways are medium-sized cattle, smaller than cattle like the Black Angus and others. The bulls can weigh anywhere from 1,800–2,100 lb (815–955 kg), while the cows usually range from 900–1,300 lb (400–600 kg) (Belted Galloway, 2022). The belted coat is definitely their standout feature. The body is encircled by a white belt which stands out against the predominantly black coat. Black is the most common color, but there are also variations like dun and, less frequently, red.

The belted pattern can be passed on from one generation to the next with the help of a single parent who carries the gene for it. It's worth noting that the defining characteristic of the belt is just the white portion. Bulls typically sport a sleek black coat and white belt, while cows may have a few subtle white markings on their lower legs and still meet registration requirements. The Belted Galloway stands out with its distinctive color pattern, which adds a lot to its overall visual charm.

Genetic Makeup and the Secrets of Hardiness

The genetic makeup of the Belted Galloway is truly remarkable. They are naturally hornless. This feature gets rid of the need for dehorning, which is a pretty standard thing in cattle management. Belted Galloways also have an unusual double coat consisting of two separate parts.

The coat has two layers—an outer layer that is long and coarse, and an inner layer that is dense and soft. The outer layer protects against rain and wind, while the inner layer keeps them warm during cold winters. This double coat helps them thrive in colder climates without gaining too much back fat, which keeps their bodies lean and their meat well-marbled.

Comparative Analysis

The Belted Galloway stands out from other popular cattle breeds with its unique set of characteristics. Let's take a closer look.

Angus

Angus cattle are famous for their top-notch beef, thanks to their Scottish heritage and distinctive black coat. However, they don't have the belted pattern and are usually bigger than Belted Galloways. Angus cattle need better feed compared to the Belted Galloway's ability to do well on less fertile pastures.

Hereford

Herefords are known for their unique white face and red body; they are a breed with horns. They can adapt to different conditions and have a reputation for having a good temperament. On the other hand, Herefords require more maintenance for their coat compared to the Belted Galloway's double coat, and their meat generally has a higher fat content.

Highland Cattle

These cool Scottish cattle have super long horns and a very shaggy coat, perfect for their home turf. Highland cattle are usually bigger than Belted Galloways and need more grazing land due to their hardy nature. Moreover, taking care of their long horns requires more attention.

Belted Galloway

The Belted Galloway is quite a fascinating breed in the world of cattle, thanks to its distinct physical features, special genetic composition, and impressive ability to adapt. Their eye-catching belted coat is just the start; their easy grazing, low-maintenance demeanor, and friendly nature make them a valuable asset for ranchers.

Behavioral Traits and Temperament

Stunning good looks are just half the story of the Belted Galloway. This extraordinary breed of cattle is a delight for seasoned ranchers thanks to its pleasant demeanor, unique behavioral patterns, and deep social structure. This section explores the social dynamics, environmental behavior, and handling tips for Belted Galloways, providing useful insights for anyone interested in learning about and engaging with this distinctive breed.

Social Structure

Belted Galloways are quite the social animals, always forming strong bonds within their herds. These groups are usually led by a dominant cow and have a matriarchal structure. This cow knows how to keep things in line when it comes to grazing, moving, and taking a break. It's all about maintaining a sense of order within the group. Calves stick by their moms during the early months, picking up on key social cues and grazing skills.

In the herd, there's a pecking order where the older cows usually have a higher rank than the younger ones. Furthermore, Belted Galloways are quite famous for their calm and friendly nature when it comes to socializing. You can often spot playful behaviors, especially among young calves. They love to chase each other, spar, and have mock fights, which helps them grow and develop.

The herd dynamic also involves the presence of the bull. Even though big bulls can get a bit territorial and aggressive with other bulls, they usually have a protective side when it comes to the cows and calves in their herd. They're always on guard, ready to protect against any potential dangers that may come their way. This includes keeping a watchful eye out for predators and unfamiliar animals, such as dogs and unknown people.

Environmental Influences

Belted Galloways can thrive in a wide range of environments. Out in the open fields, they have a way of grazing that really helps them make the most of their time. They usually graze with their heads down, slowly making their way across the field to avoid overgrazing. This foraging behavior helps to minimize waste and keep the pasture healthy.

Belted Galloways are pretty agile when it comes to tackling tough terrain like hills or forests. They're really good at navigating uneven ground, making it easy for them to reach grazing areas that other breeds might not be able to get to. Moreover, their thick double coat offers great protection against tough weather conditions, enabling them to graze comfortably even in cold temperatures or light rain.

When it's scorching outside, Belted Galloways know how to beat the heat. They'll find some shade or take a dip in mud or water to stay cool. Ranchers should make sure there are shaded areas or options for mud wallows available during hot spells.

Handling and Interaction

The easygoing nature of Belted Galloways makes them a breeze for seasoned ranch hands to work with. They're just naturally chill and laid-back, which makes them super easy to work with. However, it's still important to handle the animals properly to ensure their safety and the well-being of the rancher. Here are a few tips to keep in mind when interacting with Belted Galloways or any other cattle breeds:

- **Approach calmly and quietly:** All cattle breeds can get easily spooked by sudden movements or loud noises. Take it easy and keep things relaxed to make them feel comfortable and encourage smoother communication.

- **Establish yourself as the leader:** Although Belted Galloways don't tend to be hostile, it's important to take the lead when interacting with these cattle. Keep a relaxed stance and speak in a calm and friendly tone.

- **Respect their personal space:** Make sure to give Belted Galloways plenty of space, especially when they're with their calves. Just be careful not to get too close or startle them with any sudden movements.

- **Positive reinforcement works best:** Giving rewards for good behavior, like a gentle pat or a little treat, can help foster cooperation and establish trust between the rancher and the cattle.

- **Be aware of individual temperaments:** Every animal, including Belted Galloways, has its own unique character. Some cattle are just naturally more curious and approachable, while others tend to be a bit more cautious. Take your time and pay attention, adjusting your approach to suit each unique animal.

It's important to have a good grasp of the social structure, environmental adaptations, and behavioral traits of Belted Galloways if you want to interact successfully with these fascinating cattle. Ranchers who value a relaxed and considerate approach find working with them to be a delightful experience. By using the right methods and creating a bond based on positive reinforcement, ranchers can establish a mutually beneficial connection with their Belted Galloway herd. Belted Galloways are truly captivating creatures, with their stunning appearance and endearing nature.

The Lifespan of the Belted Galloways

Belted Galloways are known for being a long-lived cattle breed. Here's the breakdown on their lifespan and breeding age:

- **Lifespan:** This hardy breed can live a good long life, with a typical age range of 17 to 20 years or longer when given proper care (*Belted Galloway*, 2022). That's plenty of time for them to be part of your herd.

- **Breeding age:** While they can live for nearly two decades, Belted Galloways aren't ready to be moms until they're a bit more mature. They typically need to reach around 2 years old before their bodies are fully developed enough for safe breeding (*Belted Galloway Society*, n.d.).

- **Long breeding career:** The good news is that Belted Galloways are known for being reliable breeders well into their teen years, sometimes even into their 20s. That means you can get a good number of calves out of a single cow over her lifetime.

Chapter 2:

Basics of Belted Galloway Cattle Ranching

(Fence duty essentials: keeping our ranch in tip-top shape.)

Are you interested in the idea of raising your own herd of these one-of-a-kind bovines? This chapter explores the key aspects of Belted Galloway ranching. We'll look at things like fencing, feeding methods, land needs, and breeding procedures. Whether you're an experienced rancher looking to add some variety to your herd or a curious newcomer to the world of agriculture, this chapter will provide you with the essential knowledge to begin your Belted Galloway ranching journey. Jump aboard!

Essential Equipment and Infrastructure

There's something about raising Belted Galloway cattle that goes beyond their stunning appearance and easygoing temperament. They provide a one-of-a-kind ranching experience that focuses on being efficient and showing respect for the natural environment.

But before you start, it's important to have a good grasp of the necessary equipment and infrastructure to take care of your herd and make sure your operation runs smoothly. Let's examine the equipment, supplies, and ranch layout factors that are essential to a successful Belted Galloway ranching endeavor.

Essential Tools and Machinery

Even though Belted Galloways are famous for being easy to take care of, there are still some tools and machinery that are necessary for managing the herd efficiently. Here's a list of some important equipment:

- **Handling equipment:** You definitely need a sturdy cattle squeeze chute if you want to handle vaccinations, hoof trimming, or other veterinary procedures safely and efficiently. Consider investing in top-notch equipment that puts animal comfort and safety first.

- **Feeders and hay racks:** You'll need sturdy hay racks and some tough feeders for pelleted hay, grain, or concentrate supplements to make sure your animals get the nutrition they need. Think about ways to reduce waste and make cleaning a breeze.

- **Watering systems:** Making sure your herd has access to a dependable water source is crucial for their overall well-being. You have a few options to choose from for your pasture layout and water access. These include troughs, automatic waterers, or portable tanks.

- **Portable corral panels:** These panels are super handy for setting up temporary holding pens to sort, separate, or load cattle. They're a real game-changer! Go for panels that have a sturdy, reliable design that can handle the force from animals.

- **Tractor or ATV:** Having a dependable tractor or all-terrain vehicle (ATV) can be really useful for a range of tasks, like moving feed and equipment or taking care of pastures and fences.

- **Livestock trailer:** Absolutely essential for your operation is a livestock trailer, sized appropriately for your herd. This trailer will be crucial for transporting your cattle to the veterinarian, for attending shows or sales, or for transport to the butcher. Consider factors like the number of animals you typically transport at once, as well as the size of your mature Belted Galloways.

Designing and Setting Up a Functional Ranch

Aside from the tools, it's really important to have a well-designed ranch layout for effective Belted Galloway management. Here are some important things to think about.

Fencing

Having strong fencing is very important to make sure your herd stays safe and secure from any potential threats. Woven wire fencing with high-tensile wire is a commonly chosen option, due to its added strength, to keep calves from escaping. Four- or five-strand barbed wire fencing keeps larger cattle inside the pasture. Take into account the unique requirements of your area and reach out to fencing experts for their valuable insights.

Managing Your Pasture

Belted Galloways are great at grazing efficiently, and it's a good idea to use rotational grazing practices to keep the pasture healthy and avoid overgrazing. You can divide your pastures into sections or paddocks and rotate your herd regularly to give the grass a chance to regrow.

Shelter

Belted Galloways are pretty tough when it comes to cold weather, but it's a good idea to give them a three-sided shelter to keep them safe from the harsh elements, especially when they're calving. Basic pole barns or windbreaks can offer sufficient shelter for your herd.

Water Access

Make sure all pastures have access to clean, fresh water. Placing troughs or automatic waterers strategically ensures that your cattle can easily access water while grazing.

Calving Pens

Create dedicated calving pens to ensure a calm and secure space for mother cows and their newborn calves during the birthing process. These pens should be comfortable and provide convenient access for observation and assistance, if needed.

Maintenance and Care of Equipment

It's important to keep up with regular maintenance for your equipment so it lasts longer and works properly. Here are a few pointers:

- Establish a schedule for routine maintenance, such as lubricating moving parts, cleaning troughs, and inspecting for loose fasteners.
- Investing in high-quality equipment upfront can actually save you money in the long run, as you won't have to constantly replace cheaper, lower-quality tools.
- Make sure to prepare for winter by draining water lines or installing heat tape and stock tank heaters, lubricating machinery, and storing unused tools in a dry location. This will help prevent any damage from the harsh weather.

Setting up your ranch with the necessary tools and machinery—while also creating a practical layout that focuses on the health, happiness, and movement of your herd—is key to a prosperous Belted Galloway ranching venture. Keep in mind that routine equipment maintenance guarantees its operation and lowers the possibility of malfunctions or mishaps.

By prioritizing careful planning, responsible management, and proper equipment maintenance, you can ensure a fulfilling experience while raising these incredible cattle. In the upcoming chapter, we'll explore the ins and outs of raising healthy Belted Galloways—covering everything from feeding strategies to breeding practices.

Husbandry: Raising Healthy Herds

Having a successful Belted Galloway ranch involves more than just preparing your land. Taking good care of your animals is crucial to keeping them healthy and productive. What follows is an examination of the fundamentals of Belted Galloway husbandry, including methods of breeding and raising the cattle, as well as the significance of a thorough biosecurity strategy.

Breeding Practices and Lineage Preservation

Because of their innate fertility, breeding Belted Galloways doesn't call for a lot of fuss. Responsible management, however, necessitates familiarity with breeding processes and the implementation of methods for lineage preservation.

Selection

An important first step is to select breeding stock. Make sure the animals you choose are in good condition, have an appropriate body condition score (BCS), and have a belting pattern that matches the breed standard. Temperament, parental performance history (if known), and desired characteristics in offspring should all be considered.

Natural Mating

Belted Galloways have a reputation for being inherently good mothers and breeders. Natural mating, which occurs when bulls are allowed to move around in the herd, is less stressful for the animals than artificial insemination or embryo transfer.

Recordkeeping

It is crucial to keep precise records on breeding dates, ancestry, and individual animal health in order to trace lineage, detect possible genetic problems, and make educated breeding decisions down the road.

Lineage Preservation

You can join breed groups and take part in registered breeding programs if you're very interested in keeping certain Belted Galloway lines alive.

Rearing Techniques

Belted Galloways require consistent, loving care at every stage of their life cycles, from birth to adulthood. Some important methods of raising are as follows.

Calving Management

Give the mother-to-be a peaceful and quiet place to give birth in a clean calving pen. Keep a safe distance and intervene only when absolutely required while watching calving. It is essential to have the newborn calf consume its mother's colostrum during the first twenty-four hours so it can begin to develop an immune system.

Weaning

In the first six to eight months of their lives, calves learn the ins and outs of social cues and how to properly graze from their moms. For the sake of the mother and her calf, it is best to wean the calf gently. There are many ways to wean calves, but fence row weaning is the best and least stressful method. The mother and calf can see each other and touch noses, but the calf can't nurse because of the fence. Woven wire, or close spaced wood fencing, is best because it prevents the calf from getting through it.

Nutrition

Although Belted Galloways are excellent grazers, they may require additional feeds as the pasture ages and the quality decreases. As they grow older, calves can be supplemented with forage, hay pellets, and grain—but, initially, they need a high-quality beginning meal of mothers milk. For some ranchers, the only thing they ever give their cattle is grass—a practice known as "grass-fed" or "grass-finished" cattle. *The United States Department of Agriculture has recognized the author's ranch as grass-fed, meaning that their cattle do not consume any grain or grain byproducts.* If you want a precise feeding schedule for your herd, it's best to go to a vet or an animal nutritionist.

Vaccinations and Deworming

To keep your herd safe from common diseases and parasites, make sure to deworm and vaccinate them regularly. Before you decide which immunizations and dewormers are best for your furry friend, it's a good idea to talk to your vet about the local health hazards.

Hoof Trimming

For healthy feet and to keep lameness at bay, regular hoof trimming is a must. Make sure your cattle can move freely and graze productively by scheduling regular trimming visits with a professional hoof trimmer.

Livestock Biosecurity Plan

It's really important to make sure your Belted Galloways are protected from infectious diseases. A livestock biosecurity plan is important for keeping diseases at bay and ensuring the well-being of the entire herd. Here's a simple guide to creating a solid biosecurity plan:

- **Keep external contact to a minimum:** Try to avoid introducing new animals to your herd. If you happen to get some new cattle, just make sure to keep them separate for at least two weeks, and keep an eye on them for any signs of sickness before you mix them in with the rest of the herd.

- **Vehicle and equipment disinfection:** Establish procedures for the disinfection of vehicles and equipment that enter your land, with a focus on those that have come into contact with other ranches and farms that house livestock. This helps to avoid the accidental spread of germs.

- **Rodent & pest control:** Take action to keep the numbers of rodents and pests under control on your property. Pests and rodents pose a health risk to your cattle because of the diseases they carry.

- **Visitor protocols:** Make sure visitors follow the necessary protocols, like wearing clean footwear or disposable footwear covers and disinfecting their hands before interacting with your animals. This helps lower the chance of introducing any harmful pathogens from outside sources.

- **Recordkeeping:** Ensure that you keep detailed records of all immunizations, treatments, and health problems that you notice in your herd. This information is really useful for spotting disease outbreaks early on and figuring out the best ways to treat them.

Raising Belted Galloway cattle involves a lot of attention and responsibility. Knowing how to breed animals, using effective rearing methods, and having a strong biosecurity plan are all important for maintaining a healthy and successful herd.

Ensuring Health and Safety

In the picturesque countryside, where Belted Galloways roam freely like living works of art, the top priority is the welfare of both the animals and their caretakers. Raising these magnificent animals requires a sharp understanding of cattle behavior, a dedication to following safety measures, and a proactive attitude toward maintaining their health.

Ranching Practices and Protocols

It's essential to practice safe ranching when managing Belted Galloways. This provides a safe space for cattle and ranchers—reducing danger and promoting peace. Durable fences, designed to withstand the elements and energetic animals, are the first line of defense. It's a good idea to regularly check gates, fences, chutes, and handling facilities to make sure there aren't any unexpected dangers that could lead to accidents. Having good lighting in working areas is really important, especially during those early morning or late evening tasks; it can help prevent any unexpected trips or accidents. Accidents happen less frequently and things stay orderly with clear signage and designated routes for you and your bovine friends.

However, the key to safe handling is using gentle techniques. As any seasoned animal behaviorist can attest to, taking things at a relaxed and steady pace is the key to success. Animals tend to have a more positive reaction when they are handled in a calm and gentle manner. Managing herding, loading, and vet checkups becomes a breeze when everyone adopts a low-stress approach. Think of the positive effects on the herd and the ranch safety that would result from a gentle prod rather than a mad dash across the grass.

Understanding Cattle Behavior to Prevent Accidents

Understanding the thought process of your Belted Galloways is crucial for preventing mishaps and creating a secure environment on your ranch. Just like any other cattle, they have their own behaviors that are shaped by the herd, their surroundings, and their individual personalities. By getting the hang of these signals, you can stay one step ahead and prevent any issues from cropping up.

The flight zone is like an invisible bubble around each animal, and it serves as your guide. When you enter, just remember to give them some space and gently guide them by applying pressure at their point of balance. It's all about keeping things smooth and easy. Knowing how to identify signs of distress—such as tail flicking, moos of discontent, or raised heads—can help you defuse tense situations before they escalate. Ensuring there is plenty of room and avoiding overcrowding in handling areas helps to keep stress levels low and the herd content.

Regular Checks and Disease Prevention

Keeping your herd healthy and happy involves staying on top of disease prevention and scheduling regular checkups. A complete vaccine program that is tailored to the threats in your area is like putting up a wall against germs that can spread. Your best line of defense is to schedule routine examinations by a reputable veterinarian. Regular checkups, along with tests like stool analysis and blood work, can detect potential issues before they escalate. Think about it like catching a cold early—the sooner you step in, the easier it is to handle.

Biosecurity measures are just another way to keep things safe and secure. Keeping new arrivals in quarantine and limiting access to your ranch help prevent the spread of diseases. Make sure to keep the feeding and watering areas clean, and regularly remove manure to prevent parasites and other unwanted visitors. It may not be the most glamorous job, but it's absolutely necessary to ensure the health and prosperity of your herd.

A Bond Built on Care

Putting health and safety first is about more than simply following the law; it's about making the world a better place for ranchers and Belted Galloways. The cornerstones of a safe and peaceful ranch are safe ranching techniques, which are based on knowledge of how cattle behave. You can keep your Belted Galloway herd healthy and safe by doing routine checkups and taking precautions against diseases. An atmosphere of mutual respect and care may be fostered between people and these fascinating creatures through hard work, education, and adherence to best practices. Keep in mind that everyone wins when the herd is content and healthy. Put your dedication to health and safety first, and you'll see your Belted Galloways flourish.

Chapter 3:

Nutrition and Feeding

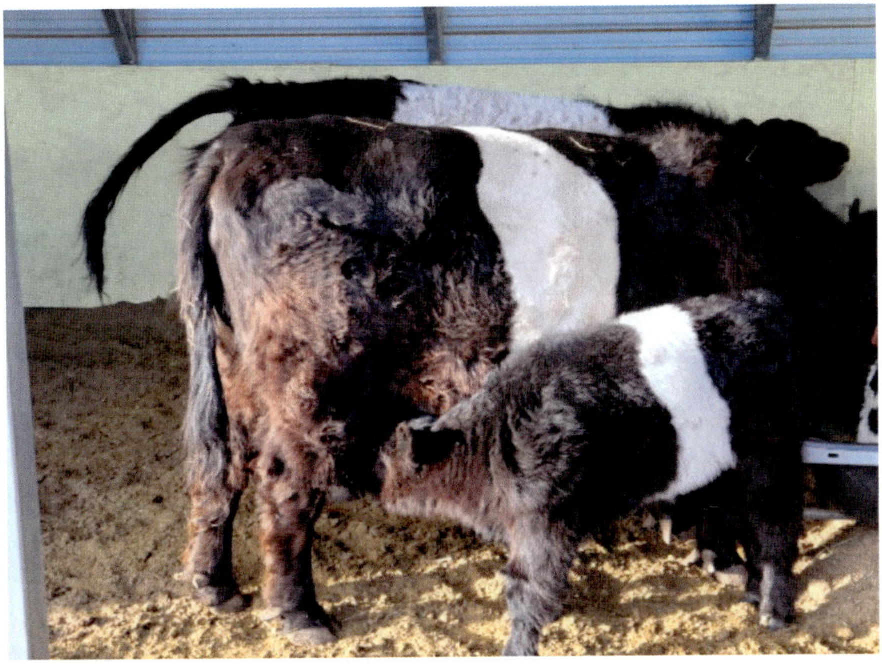

(Mooooorning! Breakfast time on the Rebel Ranch.)

These lawn mowers on four hooves are far from ordinary. Belted Galloways are great grazers, effortlessly transforming pastures into mouthwatering, marbled beef. But, like any high-performance engine, they require the proper fuel to flourish.

In this chapter, we'll explore the interesting world of Belted Galloway nutrition. We'll take a closer look at their grazing habits, discuss the importance of supplemental feeding, and provide some helpful tips for creating a personalized feeding plan for your herd. Prepare yourself to discover the secrets of ensuring your beloved cattle are content, in good health, and able to produce high-quality beef—all thanks to the importance of proper nutrition!

Dietary Needs of Belted Galloway Cattle

Belted Galloways love to munch on a diet that matches their natural roots—a diet full of fresh forage—unlike their grain-fed counterparts. These tough cows have adapted to become skilled grazers, carefully transforming pastures into the flavorful and nutritious marbled beef that people love.

We pride ourselves on our dedication to 100% grass-fed, grass-finished principles; this distinguishes us from other ranchers who may opt to supplement their cattle's diet with grain or grain byproducts. We make sure our Belted Galloways get the purest, most natural nutrition possible. There are no artificial additives or unnecessary fillers here, and all GMO-free!.

Nutritional Requirements Throughout the Lifespan

Belted Galloway's dietary requirements change as they grow older. It's important to grasp these evolving needs to make sure everything is running smoothly and efficiently.

Calves

Baby calves depend on their mom's colostrum (in the milk) during their first day of life to get important antibodies that strengthen their immune system. After this first phase, the colostrum disappears and they continue drinking mother's milk, which is packed with nutrients to help them grow quickly. As calves grow up, they start munching on grass with their mom, slowly building up their rumen. This special stomach compartment helps them break down those tough plant fibers. Around 6-8 months, calves can be weaned and start eating high-quality forage.

Growing Cattle

Calves and yearlings need a diet that's packed with protein and energy to help them grow and develop their bones. High-quality hay or pasture is essential for providing the necessary nutrients, and protein supplements—such as alfalfa and grass hay, legumes, or hay pellets—can be helpful in areas where forages lack sufficient protein. Keep an eye on how your young cattle are growing and make sure to adjust their feed intake as needed.

Pregnant and Lactating Cows

Pregnant and lactating cows have much higher nutritional needs compared to dry cows. During pregnancy, the main concern is the growth of the baby, while nursing requires extra energy for producing milk. It's really important to have more access to good forage; consider adding protein supplements during these times. It's a good idea to chat with your vet or a livestock nutritionist to come up with a feeding plan that's just right for your pregnant and lactating cows.

Mature Bulls

Adult bulls need a well-rounded diet to keep their bodies healthy and support their reproductive functions. Decent hay should do the trick, but when it's breeding season, a little extra protein might be helpful. Keep an eye on your bulls' body conditions and make any necessary adjustments to their diet.

Seasonal Adjustments

The quality of pastures naturally changes throughout the seasons. Spring pastures are so lush and full of protein and energy, while winter forages might not have as much of either. Here's what you can do to adapt your feeding strategy.

Spring and Summer

In the spring and summer months, it's important to focus on having easy access to fresh pastures. Actively growing grass has a high moisture content and nutrient profile that can meet the nutritional needs of the herd. Watch for signs that the quality of the forages is dropping as the season goes on, and add hay if needed.

Fall and Winter

During the changing seasons, the pastures tend to dry out and lose some of their nutrient content. Hay starts to play a larger role in the diet. Pick top-notch hay varieties with enough protein to meet the needs of your cattle when pastures are scarce. You might want to think about adding some protein supplements, such as canola meal or other oilseed meals, based on the advice of your veterinarian or livestock nutritionist.

Just a friendly reminder: It's a good idea to chat with your veterinarian or a livestock nutritionist to figure out the protein supplement that works best for your cattle's unique needs and any allergies they might have.

Balancing Forage and Feed

Your Belted Galloways are mostly grazers, but adding some supplemental feeding can help improve their health and productivity while still following a 100% grass-fed, grass-finished approach. Here are a few things to keep in mind:

- Make sure to always have top-notch forage available all year round. Focus on maintaining well-managed pastures and opt for hay varieties that are known for their nutritional value.

- Supplements should be chosen based on individual needs, such as addressing deficiencies or meeting increased nutritional demands during important life stages like pregnancy, lactation, or growth. Using protein supplements, mineral mixes, and even small amounts of high-quality forages like alfalfa can be helpful when used strategically. However, use caution when feeding alfalfa to cattle, as it can cause bloat.

- Try rotational grazing to make sure your animals always have access to fresh, top-notch forage all year round. This helps give the pastures a chance to recover and ensures that your herd gets the most nutritional benefit.

- It's a good idea to chat with your veterinarian or a livestock nutritionist to come up with a personalized feeding plan for your Belted Galloways. They can evaluate your unique circumstances—such as forage quality, climate, and herd needs—and suggest suitable supplements and feeding strategies to uphold your dedication to a completely grass-fed, grass-finished approach.

Fueling Excellence Through Grass

If you want to keep your Belted Galloways healthy, growing, and productive, it's important to understand their dietary needs and feed them high-quality forage throughout their lifespan. This will help them thrive. Let me tell you about the amazing benefits of the grass-fed approach.

Natural Advantage

Belted Galloways have digestive systems that are perfectly suited for a diet that consists mainly of forage. Grass is important for providing the necessary nutrients for developing a strong immune system, maintaining a healthy weight, and producing high-quality meat with a desirable marbling profile.

Environmental Responsibility

Raising grass-fed cattle has a much smaller impact on the environment when compared to grain-fed operations. When you reduce your dependence on grain, you're helping to create a more sustainable agricultural system.

Superior Taste

Research indicates that when compared to grain-fed beef, grass-fed beef has a more flavorful character and contains more healthy Omega-3 fatty acids (Gunnars, 2019). This results in a tastier and healthier product for customers.

The Art of Observation

Keep in mind that there isn't one feeding strategy that works for all Belted Galloways. The secret is to watch your herd. Keep an eye out for indications that your feeding strategy is effectively meeting their needs. Does it take care of their physical health? Are they generally energetic? Is the milk production sufficient for lactating cows? If you have any concerns, just reach out to your veterinarian or livestock nutritionist and they can help you adjust your feeding plan as needed.

The Reward of Care

When your Belted Galloways are grass-fed and grass-finished, they provide high-quality beef to eat. Adjust your feeding plan to meet their specific needs and you're not only taking care of their bodies, but also building a relationship with them based on respect for their natural dietary needs. This commitment not only leads to a healthy and thriving herd, but it also helps cattlemen continue ranching the way they do so that they can keep providing this high-quality beef for customers. So, learn how to feed your Belted Galloways as grass-fed, and watch your herd grow and thrive.

Feeding Schedules for Belted Galloways

Although Belted Galloways thrive in verdant, grassy fields, your climate poses its own particular obstacle. Depending on where you reside, fresh grass could be scarce in the spring and summer due to lack of rain or inability to irrigate sufficiently; thus, a feeding plan that takes this into account is required. Let's talk about how to keep an eye on your herd's vitals and progress so you may change their diet as needed.

Adapting to Your Climate

Due to the scarcity of fresh grass where you live, it is crucial to establish a feeding regimen that emphasizes hay pellets. To tailor a strategy to your specific needs, consider the following.

Early Morning Meal

A healthy breakfast of premium grass hay pellets is the perfect way to start the day. Pellets are a convenient way to give your Belted Galloways the nutrition they need because they are a concentrated form of food. Think about your cattle's age, weight, and activity level to figure out how much is right; your vet can give you precise advice.

Free-Choice Hay

Grass hay of high quality should be available 24 hours a day. Hay is a great source of roughage, which helps with digestion and keeps the gut healthy. It is important to select hay varieties that are both nutritious and tasty for your cattle. When looking for hay, Belted Galloways do well with orchard grass or Timothy hay.

Water

Maintain constant, unfettered availability of potable (clean) water. Digestive processes rely heavily on water, which is essential for all body functions. Check the water level in the troughs frequently and fill them up as necessary.

Adaptations for the Season

You may need to make little tweaks to your feeding plan throughout the year, even though your fundamental schedule will stay the same. To meet the increased energy needs during the colder months, you might, for instance, give more hay or pellets than usual.

Commercial Convenience vs. Organic and Homemade Options

You can buy grass hay pellets that are easy to use and provide regular nutrition. Look for names that don't have any fillers or additives, but have a lot of protein and fiber. During times when fresh food is hard to come by, hay pellets can be especially helpful.

You could, however, use organic hay or make your own hay mix, instead, if you have access to either. Pesticides and fertilizers are used less where organic hay is grown, which is good for your animals' health in other ways. You can make homemade feed mixes that are tailored to your cattle's needs, but you need to plan carefully and talk to a livestock nutritionist to make sure it contains the right amount of nutrients.

Monitoring and Adjusting Feeding Plans

A big part of figuring out how well your feeding plan works is keeping an eye on your Belted Galloways' health and growth on a regular basis. Here are some important things to keep in mind:

- **The Body Condition Score (BCS):** Use a Body Condition Scoring method to figure out how much fat your cattle have stored overall. A BCS chart can help you figure out if your cattle are too thin (needing more food), too fat (needing changes), or just the right weight.

- **Levels of energy:** Keep an eye on how active your cattle are in general. Are they tired or do they have a lot of energy to graze and move around? Less exercise can mean they are not getting enough calories.

- **Manure consistency:** The consistency of manure can tell you a lot about their gut health. If the manure is firm and well-formed, it means the diet is balanced. If the manure is loose or watery, it could mean the diet isn't balanced.

- **Condition of the coat:** A healthy coat is a good sign of general health. A dull coat with rough hair could mean that the animal isn't getting enough food.

You can change your feeding plan as needed by watching out for these signs and talking to your vet. For instance, if cows are dropping weight while they are nursing, you might want to give them more high-quality hay and grass hay pellets.

Remember that there are more ways to feed Belted Galloways successfully in your area than just setting up a standard grazing area. You need to be creative, flexible, and know a lot about your specific climate and grass. Here are some more things to think about to make your feeding plan work better:

- **Local know-how:** Talk to ranchers or agricultural extension workers in your area. They can give you useful information about feeding methods that work well in your area. Take what they've taught you and use it with your own Belted Galloway herd.

- **Hay quality testing:** You might want to check the quality of your hay every so often to make sure it meets your cattle's nutritional needs. Hay analysis tells you a lot about the protein, fiber, and other important nutrients that are in the grass, so you can decide if they need to take supplements or not.

- **Making use of technology:** Several advances in technology can help you keep an eye on your animals' health and food intake. Wearable tech can, for instance, keep track of how active the cattle are and spot possible problems early on. Automated feeding systems can also make sure that animals always have access to hay and hay pellets, which cuts down on waste and increases efficiency.

The Art of Balance: Meeting Nutritional Needs With Practical Considerations

In the end, the best time to feed your Belted Galloways depends on where you live and how convenient it is for you to do so. While store-bought feeds are convenient, looking into organic or home-made choices can give you more control over the ingredients and may have health benefits. No matter what feeding plan you choose, make sure your animals always have access to clean water, good food, and close monitoring of their health and growth.

You can make sure your Belted Galloways get the food they need to grow by learning how to feed them grass hay and pellets, adjusting your feeding methods to fit your area's climate, and staying open to new ways to improve. Remember that a healthy, well-fed herd means not only the best growth and production, but also the pleasure of seeing these beautiful animals thrive in your care.

Nutritional Management for Belted Galloways

An excellent feeding plan based on high-quality grass hay pellets and hay should be the basis of your Belted Galloways' diet, but there are times when they may need extra nutritional support. This section talks about the vitamins and minerals that Belted Galloways need, how to use supplements correctly, and how to keep track of how well your general diet plan is working.

Important Minerals and Vitamins

All cattle, including Belted Galloways, need a certain mix of vitamins and minerals to stay healthy, grow, and have calves. These are important.

Macrominerals

These minerals are essential for many biological processes and must be present in sufficient amounts. Sodium and chlorine are important for pH balance and staying hydrated, while calcium and phosphorus are important for building and maintaining bones. Muscle contractions and energy metabolism both rely on magnesium.

Microminerals

Even though smaller amounts are needed, these minerals are still very important for many things. Copper is needed to make red blood cells, and selenium helps the defense system work and keeps muscles healthy. Zinc is important for growth, breeding, and healthy hooves. Manganese is important for enzymes to work and for building bones.

Vitamins

Belted Galloways need vitamins A, D, and E to stay healthy. Vitamin A helps with reproduction, health, and eyesight. To keep bones strong, vitamin D helps the body absorb calcium. As an antioxidant, vitamin E keeps cells from getting hurt.

Tips on When and How to Take Supplements

Supplements should be used carefully to make up for specific nutritional deficiencies or meet the higher needs of growing calves during important stages of their lives. Let's discuss when it might be necessary to take supplements.

Deficiencies in Forage

The amount of nutrients in hay and pastures can change based on things like the quality of the land and how the plants were grown. If tests show that you aren't getting enough of certain vitamins or minerals, supplements can help.

Life Stage Requirements

Cows that are pregnant or nursing, as well as calves that are growing, need more food. During these important times, supplements like protein sources or extra calcium can be helpful.

Seasonal Adjustments

During the winter, when new forage is harder to come by, you can make sure your animals get the nutrients they need by giving them extra vitamins or minerals.

Delivery Methods for Supplements

There are different kinds of supplements, and each has its own benefits:

- **Mineral blends that are easy to use:** These make it easy to give various minerals in one free-choice feeder. Pick a mix that was made for cattle that are fed grass.

- **Trace mineral blocks:** These blocks give you slow-release access to zinc, copper, and selenium—which are all important trace minerals.

- **Injections:** Certain vitamins or minerals are injected, skipping the digestive system and making absorption faster. This method can be used to treat specific problems or individual animals.

- **Topical sprays:** Some vitamins, like vitamin D, can be put on the skin to help the body absorb them. Before you use this method, talk to your vet.

Working With Your Veterinarian for Optimal Nutrition

When creating a personalized diet plan for your Belted Galloways, it is essential to consult with your veterinarian. After taking into account factors like weather, forage quality, and herd requirements, they will be able to advise you on the best supplements to use or ways to modify your feeding schedule. They can also interpret blood test results and offer advice on how to fix any problems that may have been found.

Chapter 4:

Pasture Management

(Just another day in paradise: Chow time on the ranch.)

Have you ever wondered how those picture-perfect fields that are teeming with life and bright green simply show up? It's not just Mother Nature's doing, Rancher. It's the science and art of managing pasture, and if you know how to do it right, you can turn a simple piece of land into a healthy environment for your Belted Galloways.

If you want to be a real "grass whisperer," this chapter will teach you how to grow the perfect pasture—one that feeds your animals, supports wildlife, and keeps your land growing for years to come. Let's figure out how to make a Belted Galloway paradise one blade of grass at a time!

The Foundation: Soil Health

Before your Belted Galloways can eat from lush, green pastures, there's an important base that needs to be in place: healthy soil. Healthy soil is a live ecosystem full of good microbes, fungi, and earthworms that all work together to make the soil fertile for grasses to grow.

For your pastures to reach their full potential, they need the right "nutritional" setting, just like your Belted Beauties need a balanced diet. This part teaches you how to evaluate your soil, use conservation methods, and create the perfect pasture for your Belted Galloway herd.

Note: The US Department of Agriculture (USDA), Natural Resources Conservation Services (NRCS) recommends planting native grasses in pastures if they grow in your area (Darris, 2003).

Testing and Improving Soil Quality

Finding out what's going on below the surface is the first thing you need to do to get the most out of your pastures. Soil testing is very helpful because it tells you how healthy your soil is right now and where it needs work. Most of the time, a full soil test will measure:

- **pH:** The pH of soil shows how acidic or basic it is. Most plants that you want to grow do best in soil that is slightly acidic (pH 6.0–7.0). If your soil test shows a big difference from this range, you may need to add things like lime (for acidity) or sulfur (for alkalinity) to get the pH levels back to where they should be (Gibson, 2023).

- **Nutrient levels:** Phosphorus, potassium, and nitrogen are some of the important minerals that will be checked. Lack of these important nutrients can stop grass from growing. You can come up with a fertilizer plan based on the test results to fix any nutritional imbalances. Only use natural, not synthetic, fertilizer.

- **Organic matter content**: Soil that is healthy has a lot of organic matter in it, which gives plants nutrition and helps them hold on to water. If a test of your soil shows that it doesn't have enough organic matter, you can raise its levels by doing things like planting cover crops and adding compost.

Soil Conservation Techniques

Maintaining healthy soil demands the same dedication to prudent land management as tending to Belted Galloways. Presented below are a few important approaches.

Rotational Grazing

Think of your Belted Galloways as being at a buffet, but instead of one big table full of food, there are several smaller areas. The same is true for rotational grazing. You use temporary fencing to split your pastures into smaller sections, or paddocks. Your cattle feed in one paddock at a time, giving the others time to rest and grow.

When cattle have access to a large area, they tend to feed heavily on the grasses that taste the best and are easiest for them to get to, while ignoring others. Rotational grazing puts more of their grazing load on a smaller area. This makes them eat a wider range of plants and keeps them from destroying the entire landscape.

When paddocks are grazed, grass needs time to rest so it can grow new blades and roots. This makes sure that your animals always have access to high-quality food during the grazing season. When animals graze on the same area over and over, they can pack down the dirt, making it harder for water and air to move through. Rotational grazing gives the soil a chance to rest and improve.

Planned Rest Periods

Even the most dedicated buffet-goer needs a break! For your fields to really grow, you need to plan times to rest. Make sure that an area has enough time to rest and grow, after a grazing cycle, before letting your Belted Beauties back in. The best amount of time to rest relies on things like the grass type, the weather, and how hard the animals are grazing. Rest periods usually last between 30 and 60 days, but talking to an agricultural extension agent in your area can help you figure out the best time frame for your case.

Cover Cropping

Cover crops are like green mulch in nature. You use these plants when your pastures aren't being grazed. Examples of cover crops include legumes like peas and clover, brassicas like turnips and radishes, or cereals like rye and oats.

Cover crops protect the soil from wind and water runoff, especially when the ground is hard to work with because it's not growing anything. Cover crops are either grazed by your cattle or tilled back into the earth when they are fully grown. This adds valuable organic matter. This organic matter makes the soil more fertile, better at holding water, and generally more stable.

Cover crops made of legumes have a special ability—the nodules in their roots hold micro-organisms that fix nitrogen. These bacteria change nitrogen in the air into a form that plants can use. This improves the soil and lessens your need for fertilizers.

No-Till Seeding

In traditional ways of planting, the earth is often tilled to make a bed for the seeds. This makes the soil more easily worked, but it can also upset the delicate balance of the soil environment, which can cause it to wash or blow away. No-till sowing is a more environmentally friendly method.

Specialized planters make small holes in the ground where seeds can directly be placed. Over time, no-till methods help the soil build a healthy structure with good aggregation (clumping), which makes it easier for water to move through and drain without washing away. Soil that isn't disturbed is full of good microbes that help nutrients move around and keep the soil healthy.

The Impact of Soil Health on Forage Quality

When you have healthy land, your Belted Galloways will have access to good food. Here's how the condition of the soil affects the feed your cattle eat:

- **Nutrient Content:** Grass gets all the nutrients it needs for healthy growth from earth that is in good health. Grass that is high in nutrients gives your Belted Beauties a better, more balanced meal.

- **Taste:** Grass growing in healthy soil is usually more tasty for cattle, which makes them graze more effectively and efficiently.

- **Plant Diversity:** Healthy earth supports a wide range of plant species, giving your animals a better and more nutritious way to graze.

- **Drought Resistance:** Soil that is healthy and has a lot of organic matter holds on to water better. This makes plants more resistant to drought stress and gives your cattle a more stable food source through all seasons.

By putting soil health first, you're not only taking care of the base of your pastures, you're also setting up a system for grazing that will last. Healthy soil means healthy grasses, good food for your Belted Galloways, and an environment that works well for everyone. Consider yourself the "grass whisperer," test the soil strategically, and use conservation methods. Then, watch as your fields turn into a lush, green paradise that feeds your animals and protects the land for future generations.

Selecting the Perfect Grasses for Belted Galloway Paradise

Not only does healthy soil help a Belted Galloway pasture grow, but choosing the right grasses is also important. Different types of grasses with different growth patterns, tastes, and ability to handle different climates will help your animals have a varied and healthy diet. This section talks about the art of choosing grasses. It will teach you how to pick the best types for your area and temperature, even with the problems that climate change is causing.

Picking the Right Grass for Grazing

To choose the right grasses for grazing Belted Galloway cattle, you need to know a lot of things: the temperature, the type of soil, the cattle's nutritional needs, and how they will be grazing. Here's more information on each part.

Climate

The climate is a very important factor in choosing the plants that will do well in a certain area. For best growth, different types of grass need different amounts of water and warmth. For example, cool-season grasses, like Kentucky bluegrass and perennial ryegrass, do well in places with mild winters and average amounts of rain all year. Warm-season grasses, like Bermuda grass and bahiagrass, do better in hotter places where the growing season is longer and there is less water.

Type of Soil

The type of soil affects the choice of grass species because different grasses do better in various kinds of soil. Sandier soils are good for grasses like coastal panic grass and Bermuda grass because their deep roots can reach water and nutrients even when it's dry. Instead, clay soils might be good for grasses like tall fescue and orchard grass, which can grow in soils that are packed down and don't drain well.

Patterns of Grazing

It is important to know how Belted Galloway cattle graze in order to choose grass types that can handle grazing pressure and grow back quickly. Grasses that grow prostrate, like perennial ryegrass and white clover, can handle being grazed and walked on more often because they can grow back from the base even after being grazed down.

Nutritional Requirements

Grass's nutritional value changes based on things like its age, species, and how fertile the soil is. It's important to think about the nutritional needs of Belted Galloway cattle at different times of their growth—such as during pregnancy, lactation, and maintenance—when choosing grass varieties for grazing. For the health and production of the cattle herd, high-quality grasses with the right amount of protein, energy, and fiber are best.

Biodiversity and Resilience

Promoting biodiversity in pasture systems can make them more resistant to natural stresses and improve the health of the ecosystem as a whole. Adding a variety of grasses, legumes, and forbs to grazing land can have many benefits, such as increasing the production of food, fixing nitrogen in the soil, and making the soil more fertile. Interseeding and rotational feeding can help maintain a variety of species and lower the risk of problems related to monocultures, like weed growth and nutrient loss.

Seasonal Planting and Rotation

Planting at the right time of year and rotating pastures are vital parts of managing pastures for Belted Galloway cattle. These actions help preserve the right amount of forage available, improve the health of the land, and stop animals from overgrazing. Let's take a closer look at each part.

Planting for the Season

Planting different types of grass at different times of the year is important for getting the most out of food and nutrients. In places where the seasons change, cool-season grasses—like perennial ryegrass, orchard grass, and fescue—are usually planted in the fall or early spring, when the soil is cool and moist enough for the seeds to germinate and grow. These plants do best when it's cooler, and cattle have lots of food in the spring and fall when it's not as cold.

Warm-season grasses—such as Bermuda grass, bahiagrass, and switchgrass—are usually put in during late spring or early summer when the soil is warm enough to allow the plants to grow quickly. While cool-season grasses may lay dormant in the summer, these grasses thrive in hot, dry climates and continue to provide high-quality fodder.

Rotation

In rotational grazing, pastureland is split up into smaller paddocks and cattle are moved back and forth between them so that the plants can rest and grow back. Rotational grazing keeps animals from overgrazing, spreads manure evenly, and makes the land more fertile. Ranchers can make sure that cattle have access to new, high-quality forage by moving them around between different paddocks or pastures. This also gives areas that have already been grazed a chance to recover.

Rotational grazing also helps desirable grass species grow by stopping selective grazing and pushing animals to eat a wider range of forage plants. In the pasture environment, this helps keep the variety of species and its resilience.

Impact of Climate Change on Grass Variety Selection

Grassland environments are facing big problems because of climate change, and this can affect the kinds of grass that Belted Galloway cattle graze. Here are some ways that climate change might affect the choice of grass varieties.

Shifts in Growing Seasons

Climate change can affect when and how long growing seasons last, which can make some types of grass less suitable for ranching. In places where cool-season grasses usually grow well, warmer temperatures and changed patterns of rainfall may help warm-season grasses spread. Because of these changes, ranchers may need to change when they plant and how they graze their animals.

Increased Frequency of Extreme Weather Events

Extreme weather events—like droughts, storms, and heatwaves—are happening more often and with more force because of climate change. These things can hurt pastureland, change the way animals graze, and make it harder to find food. In climate-affected areas, it may become more important to choose grass types that are hardy and can handle extreme weather, like drought- or flood-tolerant species.

Changes in Pest and Disease Dynamics

Pests and diseases that harm grassland environments can move around and become more common as a result of climate change. Because of changes in the weather, some types of grass may become more vulnerable to pests or diseases. To lower the risks, ranchers may need to keep an eye out for new threats and use grass types that are resistant to pests and diseases in their grazing systems.

Techniques for Sustainable Grazing

Sustainable grazing practices are important for keeping pastures healthy, keeping the soil fertile, and making sure that grazing areas will be productive in the long term. All animals, including Belted Galloway cattle, do better in grazing systems that are well-run and put animal comfort, biodiversity, and soil health first.

Let's take a look at three important eco-friendly grazing strategies: rotational grazing, managing herd size in relation to pasture availability, and sustainable practices for the long run.

The Paddock Division

Pastures are split up into smaller paddocks with temporary fencing or natural obstacles, like hedgerows or bodies of water. Each paddock is the right size to give the herd enough food for a certain amount of time, usually a few days to a few weeks, but this depends on how much forage is available and how fast the animals grow.

Grazing Rotation

The animals are moved from one paddock to another on a well-planned routine schedule. This rotation gives areas that have been grazed a chance to rest and grow back, while still giving animals access to new, high-quality food. The amount and length of feeding are carefully controlled to keep plants from being overgrazed and to encourage plant diversity.

Rest and Recovery

Forage plants can rebuild their root systems, get more carbohydrates, and grow new leaves during breaks between grazing rounds. This makes fields healthier, more resilient, and lowers the risk of soil erosion, nutrient loss, and weed growth. By spreading manure more widely across the land, rotational grazing also improves nutrient cycling.

Rotational grazing improves forage quality by concentrating cattle in smaller areas for shorter periods of time, which stimulates selective eating of high-quality forage. This makes better use of the forage and improves the output of the animals. Rotational grazing can also help control parasites by giving internal parasites breaks in their life cycles through planned pasture rest times. Here's how it works: Animal waste, like that from cattle and sheep, contains parasite eggs. On the grassland, these eggs will hatch into larvae, the infectious stage of the animal's life cycle. We break the cycle by moving the animals to new fields before the larvae grow up. During the rest phase, the remaining larvae die off, decreasing the likelihood that animals would eat them when they graze that pasture again later.

Balancing Herd Size With Pasture Availability

Achieving the right balance between herd size and pasture availability is key to avoiding overgrazing, maintaining healthy pastures, and keeping cattle productive. How can you find this balance? Here are some ideas.

Stocking Rate Management

The stocking rate, or the number of animals per unit of land area, needs to be carefully controlled to prevent the pasture from getting worse and make sure there is enough forage. Ranchers can change the number of animals they have each season or yearly to adapt to changing conditions by keeping an eye on pasture growth rates, forage quality, and animal success.

Grazing Intensity

Changing the amount and length of grazing based on the supply of forage and the condition of the pasture can help stop overgrazing and encourage long-term use of forage. For instance, when forage grows quickly, animals may be allowed to graze more intensely so that too much mature, low-quality forage doesn't build up.

Supplementary Feeding

Giving extra food when there isn't enough forage or when pastures aren't producing much can help grazing fields handle more animals and lower the risk of overgrazing. Feeding high-quality forages, hay pellets, grains, or protein supplements can help maintain animal health and performance while preserving pasture resources.

Seasonal Adjustment

Changing the size of the herd or the way it is managed, based on the seasons, can help to ensure the supply and demand of forage stays in balance all year long. For instance, lowering the size of the herd or switching up feeding methods during the winter, when pasture growth is slow, can help prevent pastures from getting worse and make sure there is enough forage for the next grazing season.

Long-term Sustainability Practices

Along with rotational grazing and managing stocking rates, long-term sustainability practices are also needed to protect the health of the land, increase biodiversity, and lessen the damage to the environment. Here are some important things that can be done to make feeding systems last for a long time:

Soil Conservation

Soil conservation techniques—like preserving vegetation cover, preventing soil disturbance, and encouraging the buildup of organic matter—can improve the structure of the soil, the flow of water, and the cycling of nutrients. No-till ranching, cover crops, and rotational grazing are all ways to help keep the soil from washing or blowing away, from becoming compacted, and from losing its nutrients.

Biodiversity Enhancement

Pastures that are more diverse—including native grasses, legumes, and forbs—have better forage quality, improved wildlife and pollinator support, and a more resilient ecosystem. Keeping plant communities different also stops the spread of pests and diseases and gives good biologics and microorganisms a place to live.

Water Management

Using water management techniques—like putting in water troughs, creating riparian buffers, and taking care of marshes—can make the water better, stop erosion, and make wildlife habitat better. Livestock can have less of an effect on natural water bodies and lessen the likelihood of overgrazing in areas near water sources if they have access to sufficient water.

Carbon Sequestration

Grazing fields can increase soil organic matter and plant photosynthetic processes, both of which have the ability to remove large quantities of carbon dioxide from the air. Rotational grazing, using different types of forage, and reducing tillage are all techniques that can help plants grow. These can help store carbon and help fight climate change.

Lastly, sustainable grazing methods are important for protecting natural resources, keeping fields healthy and productive, and making sure animals are cared for properly. Practices like rotational grazing, keeping herd sizes in balance with pasture supply, and putting in place long-term sustainability measures help make sure that grazing systems will work for future generations. By utilizing these methods, ranchers can help protect the environment, make more money, and create strong agricultural areas.

Chapter 5:

Health Care and Veterinary Care

(Curious cattle: Our herd is always up for a closer look.)

Your Belted Galloway herd lives in a world that seems peaceful, but there is an enemy that you can't see. The Bovine Respiratory Disease Complex (BRDC) is a common, but misleading, sickness. It's not a fire that you can see from a mile away. It steals your cattle's health and productivity without you knowing.

This sneaky enemy does its best when cattle aren't paying attention—causing breathing problems, weight loss, and financial issues. But fear not, Belted Galloway guardian! You can protect yourself from this silent threat by learning about BRDC, following strategic vaccination routines, and putting strong hygiene practices first. This chapter equips you with the knowledge and tools to safeguard your herd's health and ensure their well-being. Let's fight back against BRDC and make sure your Belted Beauties can breathe easy!

Common Health Issues in Belted Galloway Cattle: Identification, Prevention, and Management

The Belted Galloway cattle are known for being tough and able to survive in a wide range of situations. They are said to be immune to most bovine diseases, but to be fair, they can get sick like any other livestock, which can impact their health and production. Ranchers can make sure that their Belted Galloway herd is healthy and happy by learning about these common health problems, taking steps to avoid them, and setting up regular health monitoring routines.

Parasitic Infections

Parasites inside and outside of cattle—like worms, lice, and ticks—can impact their health and performance. Loss of weight, lack of appetite, a rough coat of hair, and changes in behavior are all signs of parasitic infections. Regular deworming, keeping living areas clean and dry, and switching fields are all ways to keep animals from being exposed to parasites. Testing the feces and talking to a vet can help find parasitic problems and develop a good plan for getting rid of them.

Respiratory Diseases

Pathogens, environmental stresses, or being too crowded can all lead to respiratory infections like pneumonia. Some signs that cattle have a lung illness are coughing, stuffy nose, hard breathing, and eating less. Proper diet, avoiding overcrowding, being vaccinated against common respiratory diseases, and having enough ventilation are all effective preventative measures. Animals that are sick may need to be given medicines right away while being under a vet's care.

Metabolic Disorders

Changes in hormones, an unbalanced diet, or worry can all lead to metabolic disorders, like acidosis and ketosis. Some signs are losing weight, not making as much milk (in cows that are nursing), being tired, and acting in a strange way. Some things that can be done to stop this from happening are making dietary changes slowly, not giving too much grain, making sure they have access to clean water, and keeping an eye on their body condition score. To fix metabolic imbalances and improve health as a whole, nutritional products and changes to management may be needed.

Foot Problems

Wet and muddy conditions, rough terrain, or an accident can all lead to foot problems, like foot rot and being unable to walk. Problems with the feet can show up as limping, stiffness, redness, and not wanting to put weight on the affected limbs. Keeping living places clean and dry, cutting back overgrown hooves, and giving them the right hoof care are all things that can be done to stop this from happening. Antiseptic foot baths and medicines may need to be given right away to treat foot rot and stop secondary infections.

Reproductive Disorders

Infertility, abortion, and placenta retention are all reproductive diseases that can affect a herd's ability to reproduce. Sometimes, the estrus cycles aren't normal, the chances of getting pregnant drop, and the embryo dies. Prevention methods include using a good reproductive control program, keeping the cattle in good shape, and regularly checking their reproductive health. Seeking help from a vet at the right time and following the suggested breeding methods can help reduce reproductive deaths and improve breeding results.

Breed-Specific Vulnerabilities

Belted Galloway cattle may have a higher risk of developing some breed-specific diseases, notwithstanding their resiliency.

Eye Problems

Belted Galloways, especially those with white or light-colored faces, may get eye problems like pink eye (infectious cattle keratoconjunctivitis) from being in the sun and being around things that are irritating to the eyes. Shade and cover, limiting dust and fly exposure, and fly management are all preventative methods. Eye infections can be managed and problems can be avoided by finding them early and treating them with topical antibiotics or anti-inflammatory drugs.

Heat Stress

If it's hot and humid where you live, Belted Galloway cattle may be more likely to get heat stress because their coats are thick and have two layers. Animals that are under a lot of heat stress may breathe faster, drool a lot, become tired, and eat less. As a preventative step, you can give them enough shade, air flow, and access to clean, cool water. Regular brushing and shearing may also help reduce heat stress by making it easier for heat to evaporate.

Coat Conditions

It is well-known that Belted Galloways have a unique coat color and a thick, protective hair coat that keeps them warm in cold weather. But they might get skin problems, like eczema and rain scald, if they are in places that are wet and muddy. As a preventative step, keep their living areas clean and dry, groom them to get rid of mud and other debris, and give them a place to stay when the weather is bad. To treat skin irritations and keep the hair healthy, topical treatments and supportive care may be needed.

Regular Health Monitoring Routines

For early detection of possible health problems and timely intervention, it is important to set up regular health monitoring routines. Consider these important parts of a full health monitoring scheme for Belted Galloway cattle.

Clinical Evaluations

Check each animal's body condition score, coat condition, mobility, and general health by giving it regular physical exams. Look for signs of being lame, experiencing trouble breathing, having problems reproducing, or other health problems that could be happening.

Examination of the Digestive System

Test the feces of a representative sample of animals to determine the parasite load and to spot any internal parasites. Change the deworming methods based on the results of the feces tests and talk to a vet to come up with a good plan for controlling parasites.

Program for Vaccination

Start a vaccine program that is based on the diseases and health risks that are most common in your area. Talk to a vet about what shots to give Belted Galloway cattle and how often to give them, taking into account things like age, herd immunity, and exposure risk.

Checks for Reproductive Health

Regular checks for breeding soundness, pregnancy testing, and reproductive health can help you keep an eye on reproductive success and fertility rates. Find and treat any problems with reproduction or breeding right away to improve breeding success and herd output.

It is possible for ranchers to improve the health, welfare, and productivity of their Belted Galloway herd by setting up regular health tracking routines, addressing breed-specific weaknesses, and putting extra emphasis on preventative measures. Belted Galloway cattle can do well in a wide range of settings as long as they get the right food, care, and attention from veterinarians. They can also help make livestock production systems more sustainable.

Prevention and Early Detection

To keep Belted Galloway cattle healthy and happy, good herd management is all about being proactive and finding problems early. Regular veterinary check-ups, along with strong health monitoring systems and being trained in early detection procedures, can help ranchers discover and handle health issues before they deteriorate. This method ensures the herd's health and efficiency.

Importance of Regular Veterinary Check-Ups

Belted Galloway cattle require regular veterinarian examinations to ensure their well-being, productivity, and lifespan. Preventative care, early disease diagnosis, and treatment regimens adapted to each herd's unique requirements are all areas in which veterinarians excel. The importance of routine veterinary examinations is highlighted here.

Preventive Care

Preventing common diseases and health problems can be achieved through the complete health evaluations, immunizations, and personalized health management regimens that veterinarians can provide. Infectious diseases, metabolic abnormalities, and reproductive concerns can be reduced in cattle by ranchers who take preventative actions and address potentially harmful situations.

Prompt Detection of Illnesses

Having your cattle examined by a vet regularly allows for the early diagnosis of any developing health problems, allowing for faster treatment and intervention. Physical exams, diagnostic testing, and laboratory analysis can help veterinarians detect even the most minor changes in an animal's health status, allowing them to treat underlying problems before they worsen. The economic losses caused by veterinary costs and lost production can be minimized through early disease detection, which also improves treatment outcomes, decreases mortality and morbidity rates, and lowering costs.

Testing and Evaluation

Lab tests and imaging investigations are among diagnostic tools that veterinarians might use to help determine the best course of treatment for a patient. The goal of diagnostic testing is to determine the infectious disease agents, evaluate the efficacy of treatment plans, and direct management decisions in accordance with the principles of evidence-based medicine.

Guidance and Training

When it comes to herd health and management, veterinarians are invaluable tools for ranchers because of the expert advice, direction, and education they provide. Ranchers can be better equipped to handle new health issues, adopt best practices, and make educated decisions when they consult and communicate with veterinarians on a regular basis.

Training Staff in Early Detection

It's super important to train the ranch staff in early detection techniques so they can stay on top of health issues and know how to respond quickly. Here's a simple way to train your staff in early detection.

Offer Education and Training

Provide thorough training courses, seminars, and learning materials on a range of topics related to herd health, illness identification, and early detection methods. Talk about common health problems, clinical sickness indicators, safe handling and restraint methods, and reporting guidelines for health difficulties.

Hands-On Experience

Offer staff the chance to gain practical skills and hands-on experience in health assessments, physical examinations, and identifying abnormal behaviors or clinical signs. Get everyone involved and excited about monitoring the cattle's health! By actively participating and engaging in health monitoring activities, we can improve our skills and boost our confidence.

Utilize Demonstrations and Case Studies

Show examples, share stories, and use practical situations to explain important ideas, teach effective methods, and strengthen learning goals. Use visual aids, multimedia presentations, and interactive activities to make it easier to understand and remember information.

Encourage Observation and Communication

Make sure staff keeps an eye on animal behavior, appearance, and interactions every day and lets management or veterinary personnel know about any concerns or observations right away. Encourage an environment where openness, working together, and teamwork are valued to promote successful communication and finding solutions.

Give feedback and support to staff members as they work on improving their skills and confidence in early detection techniques. Acknowledge and appreciate proactive behavior, initiative, and attention to detail to strengthen positive results and foster ongoing growth.

Treatment Options and Home Remedies for Balancing Tradition and Modern Veterinary Practices

People who raise Belted Galloway cattle should know when to treat common health problems on their own and when to take their animals to a vet. Understanding common drugs and treatments, merging traditional and modern veterinary techniques, and prioritizing animal comfort are essential for herd resilience and health management.

When to Treat at Home and When to Call a Vet

Minor Injuries or Ailments

Simple first aid skills—like cleaning the wound, putting on antiseptic cream, and bandaging (if needed)—can be used to treat small injuries like cuts, scrapes, or bruises. In the same way, mild digestive problems, like bloating or diarrhea, can be treated by changing the animal's food, helping them stay hydrated, and watching to see if they get better. But if the situation gets worse or doesn't get better in a reasonable amount of time, you should get your animal to the vet right away to avoid problems.

Routine Preventive Care

Vaccinations, deworming, and hoof trimming are all examples of routine protective care that ranchers can often do without the help of a veterinarian. Following the suggested plans for immunizations, deworming, and foot care will help protect from common illnesses and health problems.

Observation and Monitoring

To find early signs of sickness or injury in both individual animals and the herd as a whole, it is important to monitor them on a regular basis. If the animal's behavior, appetite, or level of exercise changes, as well as if they cough, have a runny nose, are lame, or have abnormal discharge, you should pay attention. Talk to a vet right away if you notice any signs or changes that worry you. They will help you figure out what to do next.

Unexpected Circumstances

In an emergency—like when there is a serious accident, illness, or problem during birth—a vet should be called right away. If you think there is a life-threatening problem or aren't sure how to handle a critical situation, don't be afraid to call a veterinarian for help. Getting a professional involved right away can make treatment much more effective and reduce pain.

Common Medications and Treatments

Antibiotics

Antibiotics are often used to treat bacterial illnesses in cattle, such as those that affect the lungs, the udders, or the skin. Penicillin, tetracycline, and sulfonamides are some of the most common medicines used on cattle. Follow your vet's advice about dosage, length of treatment, and time between treatments to make sure the treatment works and lowers the risk of drug resistance.

Anthelmintics (Dewormers)

Anthelmintic drugs, also called dewormers, are used to get rid of gastrointestinal and lungworms that live inside the animal. Deworming methods may be different for various animals based on their age, how the pastures are managed, and the number of parasites that live there. To keep parasites under control, switch between dewormer classes and check the egg counts in the feces on a regular basis. This will help avoid anthelmintic resistance.

Nonsteroidal Anti-Inflammatory Drugs (NSAIDs)

Pain and swelling caused by diseases like lameness, arthritis, and mastitis can be eased with NSAIDs like flunixin meglumine (Banamine) and meloxicam. To lower the chance of side effects like stomach ulcers or kidney damage, NSAIDs should only be used when needed and with the advice of a veterinarian.

Probiotics and Electrolytes

When dealing with metabolic imbalances, dehydration, or diarrhea, it may be helpful to use electrolyte solutions and probiotics to aid with digestion and hydration. Probiotics help repair the good bacteria in the gut, and electrolyte solutions replace the minerals and fluids that the body loses when it's sick or stressed.

Wound Care Products

Wound care goods—like bandages, antiseptic sprays, and wound ointments—keep wounds clean, germ-free, and safe from becoming infected. Use mild soap and water to clean the cut well. Then, put on an antiseptic solution or ointment and cover it with a clean bandage or dressing. Keep an eye on the wound to see if it's healing or getting infected and, if it does, get help from a veterinarian.

Integrating Traditional and Modern Veterinary Practices

When traditional and modern veterinary methods are used together, they can improve the health care of cattle and encourage more holistic approaches to animal welfare. Here's how to integrate traditional and modern veterinary practices.

Traditional Herbal Remedies

Common veterinary treatments may be complemented by traditional herbal remedies and alternative therapies that help with health and happiness in animals. While chamomile, calendula, and comfrey can help wounds heal and lower inflammation, they can also be used topically to encourage wound healing.

Holistic Management Approaches

Cattle populations can be promoted to be healthier and more resilient when holistic management approaches are used. These approaches include rotational grazing, pasture management, and natural herd health measures. Focus on making the environment better, reducing stress, and encouraging natural behaviors to boost the immune system, lower the chance of disease, and strengthen the health of the herd.

Nutritional Supplements

To augment conventional veterinary care, nutritional supplements such as vitamins, minerals, and trace elements help solve nutritional deficits and promote general wellness and immune system performance. Talk to an animal nutritionist or veterinarian about making supplementation plans that are special to your herd's needs, the quality of the forage, and any nutritional imbalances.

Traditional Husbandry Practices

Traditional methods of animal husbandry—like using the right housing, handling, and hygiene—are very important for keeping the animals healthy and stopping the spread of illness. Make sure that housing is clean and has good ventilation, use biosecurity measures to stop the spread of disease as much as possible, and follow good cleanliness and sanitation rules to lower the risk of the cattle becoming sick.

Collaboration With Veterinary Professionals

It's important to foster collaboration and communication with veterinarians to leverage their expertise, experience, and assets in enhancing the well-being and health of your cattle. Work closely with veterinarians to create personalized health management plans, put preventative measures into action, and successfully deal with new health issues.

To sum up, Belted Galloway cattle treatment options and home remedies that work include knowing when to treat at home and when to call a vet, being knowledgeable about common medications and treatments, and promoting holistic approaches to animal health and welfare by integrating traditional and modern veterinary practices.

Chapter 6:

Routine Health Care

(A brand new life: One day old and full of potential.)

Just picture this: A whopping $11,000 could be at stake if you neglected even one little aspect of your Belted Galloway's regular health care. It's shocking, right? Still, every year, a great number of cattle producers are taken in by the "silent thief"—the thief of *neglect.* When you don't take your Belted Beauties in for their regular checkups, you are leaving yourself and your herd open to a host of health problems, some of which might be financially disastrous.

In this chapter, we will go over the fundamentals of regular health care for Belted Galloways, giving you the information you need to fight this silent killer. Take an active role in caring for your herd, protecting its health, and avoiding expensive problems later on.

Vaccination Schedules

For your Belted Galloway herd to stay healthy and grow, vaccinations are a must. By doing so, they are shielded from infections that could kill them, and sickness is less likely to spread throughout the herd. The exact shots needed may change based on where you live and the diseases that are common there. The best way to determine which vaccines are safest for your herd is to consult your veterinarian. This section will talk about the important shots and boosts that Belted Galloway cattle need.

Black Legs

Particularly in young calves that graze on infected pastures, the deadly muscle-affecting disease known as blackleg, caused by the bacteria *Clostridium chauvoei*, can strike. Cattle are usually vaccinated when they are two to three months old, and they get booster shots every year.

Malignant Catarrhal Fever (MCF)

There are two viruses that cause Malignant Catarrhal Fever: *bovine herpesvirus-1* (BHV-1) and *ovine herpesvirus-2* (OHV-2). This ailment can cause breathing problems, abortions, and rashes. Getting pregnant cows vaccinated before they give birth helps protect their calves.

Bovine Viral Diarrhea (BVD)

Calves can get diarrhea, respiratory issues, and abortions from the common viral infection known as bovine viral diarrhea (BVD). When pregnant cows are vaccinated in the second trimester, the calves are protected through the colostrum they produce. In some methods, the calves are vaccinated directly later on.

Parainfluenza-3 (PI-3)

The respiratory virus *parainfluenza-3* (PI-3) is part of the bovine respiratory disease complex (BRD). Some vaccines, like those that protect against PI-3 and other BRD pathogens, help make respiratory diseases less severe.

The Clostridial Diseases

Diseases like tetanus, enterotoxemia, and blackleg are caused by a group of clostridial germs. Combination vaccines that target these types are often used, starting when the calve is young and repeated as the vet recommends.

Rabies

Though not common in cattle, rabies can be fatal. Vaccination, especially in high-risk areas, may be recommended by your vet.

Booster Requirements

It is necessary to get a booster shot at regular intervals after most vaccines in order to keep up immunity. The vaccine and the animal's age determine how often booster shots are needed. Blackleg boosters, for example, are usually given once a year; whereas, PI-3 combination vaccinations may be given every four to six months.

Additional considerations include the following:

- **Pregnant cows:** To ensure the safety of both the cow and the growing calf, vaccination rules may be slightly different for pregnant cows. Talk to your vet about special suggestions for immunizing cows that are pregnant.

- **Herd health history:** If you buy cattle from another herd, find out about their vaccination history so you don't give them extra shots or leave gaps in their protection.

Tailoring Schedules to Individual Animals

Although fundamental immunizations are essential for herd health, it's worth noting that immunization schedules should be tailored to each individual animal. To help you make the best vaccine plans for your herd, let's look at some important things.

Age

When deciding how to vaccinate cattle, age is a very important factor. Calves usually get their first shots when they are very young to protect them from common diseases right away. After these first shots, cattle often get booster shots at set times to make sure they stay immune. As cattle become adults, their vaccination needs may change. To keep their immunity levels high, they may need booster shots at regular intervals.

Status of Health

Another important thing to think about when setting vaccination plans is the health of each animal. Standard vaccination procedures might not work as well on animals whose immune systems are weak because they are sick, stressed, or not getting enough food. In these situations, different vaccine formulations or different vaccination plans might be needed to provide enough protection. You can find out about any health problems, and make a vaccine plan for your cattle that is specific to its needs, by talking to your vet.

Risk of Exposure

Levels of exposure to pathogens that cause disease can vary a lot based on where you live, how you manage your herd, and the conditions of the environment. There is a chance that cattle raised in areas where certain diseases are more common will get sick more often and may need vaccines or booster shots more frequently to be fully protected. On the other hand, herds that live in low-risk places may not need as many vaccinations because they are not exposed to certain pathogens as much.

Recordkeeping for Vaccinations

Responsible herd management requires careful documentation of all vaccinations. The following details should be recorded for every animal in these records:

- **Animal identification:** Clearly identify the animal using its ear tag number or another unique identifier.

- **Vaccine administered:** Record the name of the specific vaccine administered.

- **Date of vaccination:** Document the exact date the vaccination was given.

- **Vaccination withdrawal times:** To ensure meat safety, follow animal health product withdrawal time requirements. Keep precise records of product use and follow label instructions for withdrawal times.

- **Dosage:** Include the dosage of the vaccine administered.

- **Expiration date:** Note the expiration date of the vaccine used.

- **Booster due date:** Clearly record the date when the next booster vaccination is due.

- **Vaccination site:** While not always essential, some producers choose to document the injection site for each vaccination. This can be helpful in case of any local reactions or concerns.

- **Reactions:** Record any observed reactions following vaccination, such as swelling or fever. While most reactions are mild and short-lived, documenting them helps monitor herd health and identify potential issues.

Benefits of Meticulous Records

Many important reasons warrant keeping thorough records of your cattle's vaccinations. First, by keeping track of when animals require booster vaccinations, this data can assist you in maintaining herd immunity. It helps to ensure that your herd's immunity level is elevated—which, in turn, reduces the likelihood of disease outbreaks.

You can't sell or show cattle without precise records, which brings us to our second point. For the sake of their own cattle's health, many prospective purchasers insist on seeing proof of vaccinations before purchasing an animal. In the unfortunate case of a disease epidemic, thorough immunization records can be vital for identifying the disease's origin and putting measures in place to prevent its spread.

Last but not least, you can see patterns in the health of your herd by looking at vaccination records throughout time. With this data, you may fine-tune your immunization programs to meet the changing demands of your cattle.

Fly and Parasite Control in Cattle

An integral part of cattle management is controlling parasites and flies to keep the herd healthy. Improving animal comfort, preventing disease spread, and minimizing production losses are all outcomes of effective pest management. Learn about chemical-free and natural options to fly and parasite control for cattle, as well as safe pesticide and repellent use, in this comprehensive reference to integrated pest management.

Integrated Pest Management Strategies

A method called integrated pest management (IPM) uses different methods to get rid of pests in a way that is good for the environment and doesn't use too many chemicals. Let's talk about some important parts of IPM for keeping flies and parasites away from cattle.

Cultural Controls

Using methods that cut down on pests' living and breeding areas can help lower the number of flies and parasites. To keep manure from building up, this could mean doing things like cleaning barns and fields regularly and switching where animals graze so that manure doesn't build up.

Biological Controls

Bringing in native species that eat or get rid of pests can help keep their numbers in check. For instance, some types of parasitic wasps can be used to control fly numbers by going after fly larvae that are found in manure.

Mechanical Controls

Pests can be caught or kept away with physical hurdles or traps. Mechanical ways to get rid of flies on cattle include fly nets, fly tape, and ear tags treated with insecticides. When using insecticide ear tags, it is imperative to switch the brand of insecticide yearly; otherwise, the insects can build up a tolerance to the insecticide.

Chemical Controls

Insecticides can be used to get rid of flies and other parasites when they become overly abundant. But it's important to be careful and follow the directions on the label when using insecticides to keep the environment safe and prevent resistance.

Safe Use of Insecticides and Repellents

When you use insecticides and repellents on cattle to get rid of flies and parasites, you should make sure that the animals and the surroundings are safe. To keep your cattle free of bugs, you need to use poisons and repellents that are safe and work well. Pick poisons and repellents that are made to be used on cattle only, and make sure you follow the directions on the label exactly.

When using different poisons and repellents—like pour-on solutions, sprays, or ear tags— there may be different ways to apply them and important safety rules to follow. To have the least amount of damage on the environment, make sure that you don't pollute water sources, feed, or the area around the application site. Follow the rules for getting rid of empty packages and any products that haven't been used.

Wearing protective gear—such as goggles, gloves, and a respirator—can greatly reduce the risk of exposure when working with these substances. Finally, switch to a different insecticide and repellent each year, and keep an eye on the pest population to see how well your control measures are working. This will keep the pests from becoming immune to the treatments.

Natural and Chemical-Free Alternatives

There are several natural, chemical-free ways to keep flies and parasites away from cattle that can be used by ranchers who want to avoid using chemical poisons and repellents.

Mineral Tubs With Pass-Through

A good way to control fly numbers is to give them mineral tubs with a pass-through that kills fly larvae that are growing in manure. This lowers the amount of adult flies that come out of the manure, which makes cattle less vulnerable to pests.

Biological Control Agents

There is no need to use drugs to control fly and parasite populations if you bring in natural predators or parasites of pest species. One way to do this is to let fly eggs in manure be eaten by predatory insects or nematodes.

Fly Traps and Predatory Birds

Putting up fly traps or attracting birds of prey, like swallows, can help cut down on the number of flies around cattle sites. Adult flies are caught in fly traps, and birds that eat adults flies reduce their numbers.

Diatomaceous Earth

Diatomaceous earth is a natural material made up of fossilized diatoms that can be used to get rid of flies and parasites by mechanical means. When diatomaceous earth is spread on soil or bedding areas, it breaks down insects' exoskeletons, which makes them lose water and die.

Essential Oils and Herbs

Citronella, eucalyptus, and garlic are some of the essential oils and herbs that naturally keep bugs away. You can add these to feed ingredients or put them on the animal's coat to keep flies and parasites away.

Integrated pest management strategies, safe use of insecticides and repellents, and research into natural alternatives are some of the things that cattle ranchers can do to control fly and parasite numbers while also protecting animals and the environment. Key to effective pest management in cattle operations is keeping an eye on things on a regular basis and changing control methods based on the needs of the herd and the environment.

Hoof Care of Belted Beauties

Aside from how they look, these cows need to be cared for properly, and that includes paying attention to their feet. Cattle's hooves are very important to their health and well-being because they help them move around and feel comfortable. In a simple and neutral way, let's talk about how to take care of the hooves of Belted Galloway cattle.

Regular Hoof Inspection and Trimming

Cattle need regular hoof care to avoid problems like swelling and disease, just like people need their nails trimmed regularly. The first thing that needs to be done to keep the hooves healthy is to check them often. When you check the hoof, be sure to look for any signs of damage like cracks, splits, or strange things going on with the structure.

When it comes to trimming, how often it's done depends on things like the land they live on and how fast their hooves grow. Generally, it's advisable to trim hooves every six months, if needed. But some cows may need to have their hooves trimmed more often if they have certain hoof problems or if their hooves grow quickly.

To trim feet properly, you need to be patient and follow the right steps. To prevent the cattle from getting hurt or being uncomfortable, make sure you only use clean, sharp tools made for cutting hooves. Also, make sure the hooves remain at the right angle and are balanced so they can support the animal's weight equally. Consult a hoof trimming service specializing in cattle if you don't know how to do it or need assistance.

Recognizing and Treating Hoof Diseases

As a cattle rancher, you know how worrying hoof illnesses can be. There are many potential causes, including environmental variables, bacterial or fungal infections, and even genetic predispositions. You must pay close attention to the symptoms of these diseases during your regular checks if you want to keep your beloved Belted Galloway herd healthy and mobile.

Foot Rot

This sneaky bacterial infection attacks the soft tissues between the toes, causing havoc that shows up as limping and swelling. Prompt treatment is required if foot rot symptoms, such as limping or obvious inflammation, manifest. To stop the spread of the disease, quickly isolate the animal; then, clean its hooves thoroughly, removing any damaged tissue or debris. Make sure to follow your veterinarian's advice and give your animal the necessary meds to fight off the bacteria and speed up the healing process.

Hoof Abscesses

There are painful pustules full of pus that form inside the hoof capsule. These abscesses can make your cattle very lame, which can decrease their quality of life. The first thing you should do when you find a hoof cyst is make your animal feel better. Cleaning the hurt area with care and kindness is important. Carefully remove the abscess to lower the pressure and make the blood flow better. Painkillers and certain medicines may need to be used together, in a smart way, to build a strong defense against the infection that is spreading quickly. That way, you can be sure that the cyst goes away fully and that it won't come back or cause problems for a long time.

Laminitis

This horrible condition, which is also known as "founder," causes excruciating pain and limited mobility by damaging the sensitive laminae within the hoof. Laminitis is a serious condition that requires immediate medical attention, since it is caused by poor dietary habits. Some examples of these practices include unintentionally consuming moldy feed or grazing on an excessive amount of lush grass. Getting rid of pain and swelling is the most important thing you can do to fight this strong enemy. Carefully wrap the affected area and provide soft bedding to keep their feet from getting sore from being stepped on all the time. Look at what they eat and change it, if you need to, so they don't get laminitis again.

Proper Techniques and Tools for Hoof Care

To ensure efficacy and minimize stress on the cattle, it is vital to use the correct procedures and tools when caring for their hooves. Here are some important factors to keep in mind.

Limitation and Security

Make sure the cattle are properly restrained before beginning any hoof care procedures so that they don't hurt themselves or you. To ensure the animal's safety, this may necessitate the use of a squeeze chute or other handling apparatus.

Proper Trimming Technique

Take your time and carefully remove any extra length from the hooves before moving on to more specific problems, like overgrowth or misalignment. Maintaining a well-balanced trim will help your cattle stay upright and prevent lameness.

High-Quality Tools

To achieve precise results, while avoiding discomfort for the cattle, it is necessary to invest in high-quality hoof trimming gear. Having hoof trimmers, rasps, and knives that are sharp and well-maintained is vital for good hoof care.

Proper Cleanliness

When caring for hooves, it's crucial to keep practicing proper cleanliness to avoid spreading illnesses. Maintain a clean and clutter-free work area at all times, and disinfect tools after each use.

The Belted Galloway cattle under your care will be able to keep their hooves and health in good shape if you follow these guidelines. An important part of caring for cattle is checking their hooves often, trimming them when needed, and treating hoof infections proactively. These lovely creatures can live longer and be more productive with the correct information and technique for hoof care.

Chapter 7:

Emergency Care

(Sunrise on the Rebel Ranch welcomes a new arrival: Ten minutes old and ready to explore.)

You may not know this, but most Belted Galloway have health problems at least once in their lives. Whether it's a sudden illness, a problem during birth, or an injury in the field, these beautiful animals often find themselves in dangerous situations that need instant help.

As a responsible cattle owner, being ready for these kinds of events is not only smart, but it could mean the difference between life and death for your herd. This chapter goes into detail about the most important tactics and strategies you need to know to handle situations and keep your Belted Galloways safe.

Recognizing Emergencies

No matter how long you've been ranching or how new you are to taking care of cattle, learning how to act quickly when an animal is in trouble or sick can save its life. This complete guide will talk about the main signs of problems in Belted Galloways, what to do right away in different situations, and how important it is to keep things calm and safe during these very important times.

Identifying Signs of Distress or Illness

Every animal has a way of telling you when something is wrong, and Belted Galloway cattle are no different. You must pay close attention to their actions and health status. Keep an eye out for these usual signs of trouble or illness.

Behavioral Changes:

- **Agitation or restlessness:** If your usually calm cattle seem antsy by pacing, or lying down and getting back up over and over, it could mean they are in pain or discomfort.

- **Isolation:** Cattle that stay away from the rest of the group may be sick. Be wary of the one who goes off the rails. Cows getting ready to calve will seclude themselves away from the herd a few days before calving. If you're planning to move them into a calving area, this is a good indication they are close. Also, if you have animals that are getting close to giving birth and you don't see them, search the pasture for them and you may be surprised with a new calf.

- **Unusual noises:** Too much mooing or other strange sounds can be a sign of pain or distress.

Physical Symptoms:

- **Lameness:** Limping or favoring one leg could be a sign of an accident or problem with the foot.
- **Abnormal posture:** Cattle that stand with their heads dropped, backs hunched, or necks arched may be in pain.
- **Labored breathing:** Coughing, hacking, or breathing quickly or shallowly may be signs of lung problems or pneumonia.
- **Changes in eating or drinking habits:** Refusing to eat or drink, or suddenly drinking more or less than usual, can be a sign of sickness.

Immediate Actions to Take in Different Scenarios

Scenario 1: Injury

Approaching a wounded companion with calm and caution is of utmost importance. Keeping a level head and putting the well-being of the injured animal and yourself *first* is the main order of business. Staying at a safe distance will allow you to thoroughly assess the severity of the injury without further frightening the animal. You can tell if the injury is mild or needs veterinarian treatment right away with this initial evaluation.

Take immediate action to prevent consequences if the damage seems minor, like a superficial wound. To avoid infection, cleanse the wound completely using an antiseptic solution. An additional layer of protection against infection can be provided by bandaging the wound, if that becomes required. Nevertheless, it is crucial that you promptly consult a veterinarian in the event of more serious injuries, such as deep wounds or fractures.

Importantly, after you've taken stock of the situation and started treating the wound, you must separate the wounded animal from the remainder of the herd. To reduce the likelihood of additional harm coming to the wounded animal and to protect other cattle in the herd, it is best to relocate them to an isolation area. Quarantining the injured animal also makes it easier to watch its status closely, which means it may get more targeted care.

The injured animal must be closely monitored during the treatment to ensure its safety. Their well-being and speedy recovery depend on you providing them with clean water and a cozy place to relax. Keeping a close eye on their status allows you to respond quickly in case anything changes or worsens. Paying close attention to the injured animal and attending to its requirements can greatly improve their comfort and safety while recuperating, or if they have to wait for professional help.

Scenario 2: Illness

Detecting signs of sickness in your Belted Beauties demands a discerning eye and swift action. The first crucial step is to keenly observe any symptoms exhibited by the affected animal, ranging from subtle changes in behavior to more overt signs of distress. Symptoms such as fever, diarrhea, or unusual sluggishness are common indicators of illness in cattle and should be promptly noted.

To prevent the potential spread of illness within your herd, immediate isolation of the sick animal is essential. By segregating the affected individual in a separate area, you minimize the risk of contagion and protect the health of the rest of your cattle. This isolation also facilitates closer monitoring of the sick animal's condition, enabling you to provide focused care and attention.

While awaiting professional assistance from a veterinarian, it's crucial to provide supportive care to the sick cow. This includes ensuring they have access to plenty of fresh water and a nutritious diet tailored to their specific needs. A balanced and nourishing diet helps bolster their immune system and aids in their recovery process. Additionally, maintaining a warm and comfortable environment is vital, particularly during inclement weather or colder seasons, as it promotes healing and comfort for the sick animal.

Above all, timely communication with a veterinarian is paramount. As soon as you suspect that one of your cattle is unwell, contacting a veterinarian should be your immediate priority. Veterinarians possess the expertise to accurately diagnose the illness and prescribe appropriate treatment measures tailored to the specific needs of your cattle. Whether it's administering medication, providing supportive therapies, or recommending dietary adjustments—veterinarians play a pivotal role in restoring your cattle's health and vitality.

Scenario 3: Birthing Complications

(Mom helping newborn calf stand up)

If you see signs that a cow is having trouble giving birth, stay calm and carefully evaluate the situation. As a good caretaker, you need to keep your cool and pay attention to what both the cow and the calf need.

While keeping a close eye on the cow, look for any signs that the calf is in trouble, such as labor that lasts longer than usual or the calf being placed in an unusual way in the birth canal. These signs could mean that there are problems that need to be fixed to make sure the delivery is safe.

If it looks like the cow can't birth the calf on her own, something needs to be done right away. Step in to help with the birthing process. Clean your hands, and remove your watch and jewelry. Shoulder-length disposable gloves should be worn to prevent contamination and keep your arms clean (available at feed stores and others). Help the cow ease the calf out of the birth canal by applying light traction. Be careful not to hurt either the cow or the calf.

After the birth of the calf, it is essential to closely observe their well-being. The best way to keep a cow and her calf healthy is to provide them with a dry, clean place to stay. Newborns can be quickly stabilized if caregivers keep a careful eye out for symptoms of distress, such as trouble breathing or insufficient suctioning.

Postpartum problems can develop even with meticulous management. Should you experience excessive blood, a retained placenta, or any other postpartum problem—you should get medical help right away. Veterinarians have the knowledge and tools to correctly identify and treat birthing problems, which protects both the cow and the calf's health and well-being.

First Aid and When to Call a Vet

Cattle first aid is an essential part of good animal husbandry. A well-stocked first aid kit and knowing when to take your cattle to the vet are important parts of handling cattle, whether you're dealing with minor injuries or more serious health problems. Here, we will go over the fundamentals of first aid, what should be in a cattle first aid kit, and how to tell whether a veterinarian needs to be involved.

Essential Items in a Cattle First Aid Kit

If you own cattle, you know how important it is to have a first aid kit on hand in case of any unexpected medical emergencies. Make sure you incorporate the following vital items:

1. **Antiseptic solution:** for cleaning wounds and preventing infection

2. **Sterile gauze pads and bandages:** to dress wounds and provide support

3. **Thermometer:** for monitoring body temperature, especially during illness

4. **Scissors and tweezers:** for cutting bandages and removing foreign objects from wounds

5. **Disposable gloves:** to maintain hygiene and prevent cross-contamination

6. **Hoof trimmers:** for addressing minor hoof issues and preventing lameness

7. **Veterinary wound spray:** for treating wounds and promoting healing

8. **Electrolyte solution:** for rehydration in cases of dehydration or illness

9. **Sterile syringes and needles:** for administering medications or fluids

10. **Veterinary contact information:** including the phone number of your veterinarian for emergencies

11. **Additional items:** Recommended by your veterinarian.

Understanding When Professional Veterinary Care Is Needed

While simple first aid can take care of small injuries, it is important to know when to see a veterinarian to ensure the health of your Belted Galloway herd. In the event of serious injuries—such as deep cuts, fractured bones, or significant bleeding—it is imperative that you promptly consult a medical expert. Veterinary care is necessary in cases of labor complications, as well.

Fever, diarrhea, or difficulty breathing—that does not go away after a certain amount of time—may point to a more serious problem. It is best to consult a veterinarian for a diagnosis and treatment plan. A veterinarian should be consulted for the diagnosis and treatment of reproductive issues like infertility, abortion, or complications during calving. Last but not least, make sure your cattle get veterinarian help right away if they experience any unexpected or serious health problems like bloating, colic, or acute lameness.

In conclusion, as a responsible cattle supervisor, you must know how to properly administer first aid, keep a well-stocked first aid kit on hand, and know when to seek professional veterinary care. The health and lifespan of your herd depend on your readiness to anticipate and proactively handle any health issues that may arise.

Chapter 8:

Breeding and Genetics

(Rising Star: This 14-month-old red bull is a future herd leader.)

Are you aware that cattle species' genetic composition has been significantly changed through selective breeding, resulting in remarkable gains in performance and productivity? Genetic selection has increased the average daily weight gain of beef cattle by more than 30% and doubled the milk production of dairy cows in the last several decades, according to recent studies (Oltenacu & Broom, 2010).

These shocking numbers show how much power breeders and geneticists have as they change the future of agriculture by figuring out the complicated paths of heredity. Come along with us as we explore the fascinating worlds of breeding and genetics, where changing genes and trying to make everything right come together to change the ways that livestock ranching can be done.

The Genetic Canvas of Belted Galloways

Having a belted coat and a strong body, Belted Galloway cattle are a great example of how genetics and breeding can work together to shape animal breeds. Let's learn more about genetics, the management of genetic diversity, and the important part genetics plays in breeding programs.

Understanding Genetic Traits and Heredity

Belted Galloway genetics is based on the idea that traits are passed down through genes. When genes are passed from parents to children, traits are passed down through the generations. These traits include a lot of different physical and behavioral traits, such as hair color, body size, milk production, and resistance to disease.

The ideas behind heredity, the way these genetic features are passed down from one generation to the next, were first proposed by Gregor Mendel, who is known as the "father of modern genetics" (Miko, 2014). Mendel's rules of segregation and independent assortment explain how traits are passed down by genes and how they work together to create phenotypes that can be seen in cattle and other animals.

Belted Galloway breeders need to know how desirable traits are passed down through genes in order to use selective breeding plans that improve the quality and performance of their herds. By carefully choosing which animals to breed, breeders try to keep traits like the unique belted coat pattern, strong mother instincts, and good feed conversion alive.

Managing Genetic Diversity Within a Herd

It's extremely significant to keep the genetic diversity in a Belted Galloway herd intact so that the breed can stay strong and adaptable to different environmental conditions and market needs. Genetic diversity is all about the different genes in a population. It's what allows for natural selection and evolution to happen.

Mating closely related individuals, known as inbreeding, can have a negative impact on genetic diversity. It can lead to an increase in harmful recessive alleles and a decrease in overall genetic variability. Breeders of Belted Galloways use methods like outcrossing, which involves crossing unrelated individuals to inject fresh genetic material into the population, to reduce the likelihood of inbreeding depression and genetic abnormalities.

Moreover, breeders can make use of genetic evaluation tools like Estimated Breeding Values (EBVs) and genomic selection to identify animals with superior genetics for breeding, while also reducing the chances of passing on harmful genetic traits.

The Role of Genetics in Breeding Programs

Genetics are the most important part of successful breeding plans that try to make Belted Galloway cattle more productive, profitable, and long-lived. When breeders carefully choose breeding stock based on desired genetic traits and performance signs, they can speed up genetic progress and better meet their breeding goals.

The idea of genetic improvement is at the heart of breeding programs. This means choosing cattle with better genes for traits that are economically important and adding them to the breeding pool. Some of the traits that may be improved are growth rate, carcass quality, fertility, and protection from disease.

The spread of desirable genetic traits within a breed is made easier by improvements in reproductive tools like artificial insemination (AI) and embryo transfer (ET). These technologies let breeders get genetic material from exceptional cattle no matter where they live. This increases genetic diversity and raises the overall genetic merit of the breed.

Navigating Calving Season

On the ranch, the calving season is a watershed moment because it heralds the birth of a new life and the start of a long journey of care, development, and growth. Preparing for this crucial period with your Belted Galloway cattle involves a mix of anticipation, planning, and some good old-fashioned know-how. Let's dive into the important steps for managing calving season, taking care of newborn calves, and closely monitoring their health and development.

Understanding Calving

Learning how a cow gives birth is the first and most important step. Although giving birth, or caving, is a perfectly normal process, it's smart to be ready for anything. You can tell when your cow is about to give birth by looking out for symptoms including restlessness, separation from the herd, larger udder and other physical indicators.

Nutritional Needs

The dietary requirements of your cows will vary after they give birth. To keep themselves and their developing calves nourished, they need more food and energy. To assist your cows and their calves to stay healthy, make sure they have access to high-quality pasture and maybe add a mineral mix made specifically for pregnant and nursing cows to their food.

Comfortable Environment

The most important thing is to make sure the calving area is safe and comfortable. Cows have less stress and fewer difficulties when given clean, dry bedding in a quiet, undisturbed place to give birth. To further ensure their safety and make monitoring easier, you can think about setting up calving pens or shelters with wireless cameras, where you can easily observe the cattle from the comfort of your home, at any time.

Monitoring and Assistance

Although the majority of cows will give birth without any problems, you must monitor your herd closely when they give birth. You can swiftly intervene if help is required if you check regularly, particularly at night. In case you ever need to assist with calving, it's wise to have a calving kit ready with the necessary supplies, such as lubricant, obstetrical chains, and clean towels. It's a good idea to also have a warm jacket handy. If they calve during the fall and early winter months, it seems like they pick the coldest and nastiest weather to calve.

Post-Calving Care

There is a need for post-calving care for the mother cow and her newborn calf. Colostrum is the first milk a cow produces, therefore it's crucial that the newborn calf gets some of the nutrients and antibodies it needs soon after birth. Also, keep an eye out for any symptoms of trouble nursing or sickness in the mother and calf, and be ready to take them to the vet if needed. In late fall or early winter, have a blanket handy to warm a cold calf. If needed, this could save its life!

Navel Care

One of the leading causes of illness and death in newborn calves is navel infections, which can be prevented with proper navel care. A simple way to disinfect the umbilical cord stump and lessen the likelihood of infection is to dip the calf's navel into an antiseptic solution like iodine or chlorhexidine right after birth.

Being vigilant and proactive is key when getting your Belted Galloway cattle ready for calving season. A successful calving season on your ranch is within your reach if you prepare yourself by learning about the process, feeding your cows what they need, making sure they are comfortable, keeping an eye out for labor symptoms, and taking care of them after the birth.

Monitoring the Health and Development of Calves

If you want to catch problems early on with your calves, you need to examine their health and development often. The best way to keep an eye on calves is as follows:

- **Daily observation:** The best way to catch problems like lethargy, diarrhea, or respiratory issues in calves early is to visually inspect them every day. A calf's behavior, hunger, and fecal consistency can tell you a lot about their health.

- **Tracking weight gain:** Keeping tabs on the weight of calves over time allows one to evaluate their nutritional state and growth, which in turn allows one to make necessary changes to their feeding schedule. If you keep track of the weight of your calves and weigh them at regular intervals, you can see how they're developing and see any changes that aren't normal.

- **Annual veterinary exams:** The best way to monitor a calf's health and take preventative actions is to schedule wellness examinations and checkups with the vet at regular intervals. To guarantee the best possible health and performance from calves, veterinarians recommend annual checkups, immunizations, and parasite control measures.

Advancing Reproductive Technologies

Modern reproductive techniques, especially embryo transfer (ET) and artificial insemination (AI), have completely altered the cattle breeding industry. When it comes to improving herd productivity, achieving breeding goals with precision and efficiency, and enhancing genetic development, these techniques provide ranchers with unique prospects. Here, we'll discuss artificial insemination, embryo transfer, and breeding data management to maximize the efficacy of modern reproductive technologies.

Techniques and Benefits of Artificial Insemination

Artificial insemination is the process of using special tools and methods to get semen from a carefully chosen bull into a female cow's reproductive system. This method is better than normal mating in a number of ways, such as:

Genetic Improvement

AI makes it possible for ranchers to get better genetics from top sires, no matter where they live. Ranchers can speed up genetic progress in their herds and better meet their breeding goals by using the sperm of genetically superior bulls that have features like high milk production, better fertility, and resistance to disease.

Disease Control

Because AI doesn't involve direct physical contact between animals during breeding, it lowers the risk of disease transmission that comes with natural mating. This helps stop the spread of reproductive and Sexually Transmitted Disease (STD), which is good for the health of the herd and for protection.

Increased Reproductive Efficiency

AI helps ranchers improve breeding management and synchronization routines so that reproduction works as well as possible. Ranchers can increase the chances of conception and shorten the time between calves by carefully timing insemination based on the cow's estrous cycle and fertility level. This leads to more consistent breeding outcomes and better productivity.

Cost-effectiveness

It may cost more to buy AI tools and sperm at first than to mate naturally, but the long-term benefits of better genes and more efficient reproduction are worth it. AI is an inexpensive way for ranchers and people who don't have a lot of breeding room or resources to get access to high-quality genetics and make their herds perform better.

The Process and Ethics of Embryo Transfer

The process of embryo transfer (ET) includes taking embryos from genetically better donor cows, changing them, and then transferring them to recipient cows so that they can get pregnant and give birth. This cutting-edge method of reproduction has many benefits, such as:

Multiplication of Elite Genetics

ET lets ranchers quickly increase the genetic potential of their best cows by having multiple babies from a single source. This speeds up genetic progress in the herd and makes it easier for desired traits to be passed down to future generations.

Maximization of Reproductive Potential

ET lets ranchers get the most out of the reproductive potential of donor cows by getting eggs from animals with great genetics, reproductive histories, and track records. Because of this, important genetic resources can be kept and passed on within the herd.

Flexibility in Breeding Programs

ET gives ranchers more options in their breeding plans by giving them different ways to choose which genes to use and how to spread them. Ranchers can make breeding plans that meet their specific needs by mixing ET with AI and natural mating. This is useful whether they want to increase production, improve shape, or make animals more resistant to disease.

Even though embryo transfer has many benefits, it also brings up ethical questions about the well-being of both the donor and recipient animals. There are also worries about genetic diversity and the chance of unintended effects. Ethical standards and best practices stress how important it is to protect the welfare and well-being of animals used in ET procedures. This includes making sure they have the right housing, food, and medical care during the process.

Recordkeeping and Management of Breeding Data

Optimizing the efficacy and performance of embryo transfer and artificial insemination operations requires precise documentation and administration of breeding data. Full records of breeding give ranchers useful information about how well their animals are doing, how well they reproduce, and how genetic progress is being made. This helps them make smart decisions and get the best results from their breeding.

Some important parts of managing breeding data are:

- **Tracking and identification:** To record breeding results, family trees, and reproductive efficiency through time—precise and trustworthy animal identification is required. Ranchers can keep accurate records and make sure animals can be tracked throughout the breeding process by using ear tags, computer identification systems, or blockchain technology.

- **Reproductive monitoring:** Regularly checking and writing down reproductive events—like finding estrus, insemination dates, pregnancy test results, and calving records—helps keep track of fertility trends, find breeding problems, and see how well breeding practices are working.

- **Genetic evaluation:** Using tools for genetic evaluation—like estimated breeding values (EBVs), genomic selection, and pedigree analysis in breeding data management—helps ranchers figure out which animals are genetically superior, make smart breeding choices, and set selection criteria for desired traits.

- **Breeding program optimization:** By looking at breeding data and performance metrics, ranchers can make breeding programs more effective, use focused selection strategies, and reach their breeding goals. Ranchers can make the best use of their resources, reduce the number of animals that are unable to reproduce, and increase the genetic diversity of their herds by finding trends, patterns, and areas that can be improved.

To sum up, embryo transfer and artificial insemination are potent instruments for managing bovine breeding programs, increasing herd productivity, and furthering genetic advancement. Understanding how these reproductive technologies work and what they can do for ranchers, along with following moral guidelines and using good recordkeeping and data management techniques, will help them make their businesses more profitable and sustainable for future generations.

Chapter 9:

Handling and Care

(Black Beauty: A powerful two-year-old black bull keeps the watch.)

It is surprising to discover that ranches across the world are dealing with financial and ethical issues due to the mishandling of cattle. Recent research has demonstrated that animals suffer from maltreatment while being transported and handled, which results in stress-related incidents, costing millions of dollars every year (Cockram, 2019).

This alarming fact underscores the importance of taking proper care of cattle to ensure their well-being and productivity. It is vital to adhere to basic guidelines for good handling and care, such as preventing stress in ranch animals, promoting compassionate treatment, and enhancing their health and performance. By following these guidelines, we can make a positive difference in the lives of these animals and the ranching industry as a whole.

Strategies for Handling and Moving Cattle Safely

The ranch's efficiency and the animals' well-being depend on careful handling and transportation of cattle. Let us walk you through the steps of stress-free cattle handling, how to build facilities that allow for safe and efficient movement, and why it's crucial to educate staff and helpers in safe handling procedures to minimize risks and maximize outcomes.

Techniques for Stress-Free Cattle Handling

Low-Stress Handling Methods

Cattle must be handled with care to ensure they remain cooperative and calm, which in turn reduces the likelihood of harm and increases handling efficiency. Here are a few important methods.

Flight Zones

Handlers can move and approach cattle with little stress if they understand the idea of the flight zone, which is the animal's personal space. Cattle can be led in the right direction without fear or resistance if handlers stay within the flight zone's limits and use light pressure when needed.

Pressure and Release

When you want an animal to listen, you can use the pressure and release principle, which entails providing a little pressure and then releasing it when you get the desired response. Cattle handlers may communicate successfully with their animals and promote cooperative behavior, without employing force or coercion, by using subtle cues and releasing pressure at the right times.

Quiet Handling Environment

The best way to keep cattle from being anxious or stressed out while handling them is to make sure the handling area is peaceful and quiet.

Avoiding Loud Noises

Screaming, pounding, or clanging can frighten and upset cattle, which in turn makes them more stressed and less easy to handle. Keeping noise levels low in handling facilities and the areas around them helps create a calm setting where people are more likely to be helpful and calm.

Limiting Sudden Movements

The anxiety and panic reactions in cattle can cause them to act erratically and even injure themselves or their handlers if there is a sudden change in direction or movement. Cattle benefit from handling practices that are controlled and predictable because it makes them feel safe and secure, which in turn reduces stress and builds trust.

Eliminating Unnecessary Distractions

Cattle are better able to concentrate on the work at hand and exhibit less stress-induced reactions when they are not too stimulated or when they are not among other animals. A sense of peace and improved handling efficiency can be achieved by keeping handling spaces free of excessive clutter and keeping them clean and tidy.

Gentle Handling

Cattle and handlers can develop a strong bond through the use of gentle handling techniques, which in turn reduces resistance to handling procedures and increases collaboration.

Patience and Respect

By approaching cattle with kindness, understanding, and tolerance, you can earn their trust and confidence, promoting cooperative behavior and lessening the chances of stress reactions. Cattle handlers can better meet the demands of their animals by getting to know them via careful observation and study of their behavior and temperament.

Avoiding Rough Handling

Cattle can become aggressive and even hurt if handled roughly, which includes actions like pushing, dragging, or jerking. Cattle are more comfortable and relaxed when handled gently, which helps them adjust to handling facilities and transportation without incident.

Minimizing Prodding and Aggressive Behavior

Cattle might become even more stressed out and lose trust in their handlers if they are constantly being prodded, poked, or aggressively handled. To promote friendly and cooperative handling, it is best to use subtle forms of persuasion, positive reinforcement, and open communication to bring about the needed changes in behavior with little outward signs of aggression.

Proper Positioning

If you want to communicate and keep your cattle under control while handling them, you need to posture yourself correctly. With a firm grasp of the ins and outs of the flight zone and point of balance, cattle can be safely and calmly led through various transportation and handling environments.

Designing Facilities for Safe and Efficient Movement

The layout of the facilities is the first step in guaranteeing the safe and effective handling of cattle. The animals should be able to move freely and easily across a well-planned arrangement. A number of design elements—such as winding alleyways, gentle turns, and open sight lines—make this possible so that cattle can see their destination.

Cattle handlers are more likely to sustain injuries in damp environments, so it is imperative that floors and walkways be equipped with slip-resistant surfaces such as textured concrete or rubber mats. Proper ventilation and lighting are of utmost importance. Assuring adequate ventilation and good lighting both work together to create a more peaceful and healthy atmosphere by reducing the impact of shadows and other unwelcome sights.

Lastly, include safety features—like railings, gates that are placed strategically, and well-marked emergency exits—so that you can respond quickly in case of an emergency. All handling equipment and these characteristics should be inspected and serviced regularly to detect and eliminate hazards before they endanger both you and your cattle's safety.

Training Staff and Helpers in Safe Handling Practices

Education and Training

To guarantee that staff and assistants are competent in the safe and effective handling of cattle, it is necessary to provide them with thorough education and training programs. The best way to ensure that handlers are prepared for any situation is to provide them with comprehensive training that addresses issues including animal behavior, low-stress handling techniques, facility operation, and emergency protocols. It is strongly advised that ranchers of cattle participate in the

Beef Quality Assurance (BQA) program's in-person or online training in order to obtain BQA certifications. The BQA website can be found at bqa.org as of this writing (n.d.).

Hands-On Experience

Cattle handlers can hone their abilities and gain self-assurance through opportunities for practical training and hands-on experience, which enable them to apply classroom knowledge in real-world contexts. If you want to learn and grow constantly, a supervised practice session or mentorship program is a great place to start.

Communication and Collaboration

Coordination of handling operations, and the safety of both handlers and cattle, depend on good communication and teamwork among employees and assistants. The handling team will work together more effectively and create a safer environment by outlining everyone's specific duties and following established protocols.

Continual Improvement

To keep handling procedures up-to-date and in line with industry standards and best practices in animal welfare, it is important to encourage staff and assistants to continue learning and developing professionally through frequent training updates, refresher courses, and data sharing meetings. The best way for cattle handlers to ensure the well-being and safety of the cattle in their care is to set a goal of constantly improving their skills and knowledge.

Ultimately, for the purpose of minimizing stress, maximizing welfare, and achieving optimal results when dealing with and transporting cattle, it is necessary to employ a mix of effective tactics, properly planned facilities, and competent handlers. Cattle can be raised in an atmosphere that promotes their well-being and productivity when ranchers use techniques that reduce stress during handling, build facilities with efficiency and safety in mind, and offer extensive training and support to staff and helpers.

Ensuring Shelter and Comfort for Livestock

Maintaining the health, well-being, and productivity of livestock all year round requires providing them with suitable shelter and comfort. Ranchers can make a setting that is good for the animals' health by building shelters that protect them from the weather, making sure the animals are comfortable to lower their stress, and keeping the living areas clean to encourage good hygiene.

Designing Appropriate Shelters for Different Weather Conditions

When it comes to providing shelter for your livestock, one size does not fit all. Just as humans seek refuge from the elements, animals, too, require tailored protection to thrive in varying climates. From scorching summers to frosty winters, understanding how to design shelters that cater to different weather conditions is paramount in ensuring the well-being and comfort of your livestock.

Shelter Types

Selecting the appropriate shelter type depends on factors such as climate, herd size, and available resources. Common shelter options include barns, sheds, loafing areas, and windbreaks. Each structure serves a specific purpose in providing protection from adverse weather conditions such as rain, wind, snow, and extreme temperatures.

Climate Considerations

Designing shelters to withstand local climate conditions is crucial for ensuring the comfort and well-being of livestock. Features such as insulated walls, proper ventilation, and adequate drainage help regulate temperature and moisture levels inside the shelter, reducing the risk of heat stress, hypothermia, and respiratory issues.

Seasonal Adaptations

Adapting shelters to seasonal changes allows ranchers to address specific challenges associated with various weather conditions. For example, providing shade and ventilation during hot summer months helps prevent heat stress, while adding bedding and windbreaks during winter months offers protection from cold temperatures and wind chill.

Ensuring Comfort and Reducing Stress

Comfortable and stress-free livestock are not just happier animals; they're also healthier and more productive. As a rancher, one of your primary responsibilities is to create an environment where your animals can thrive, free from unnecessary discomfort and anxiety. Understanding the factors that contribute to their well-being and implementing strategies to reduce stress are essential for promoting optimal health and performance in your livestock.

Bedding Material

Choosing appropriate bedding material is essential for ensuring comfort and reducing stress in livestock. Options such as straw, hay, wood shavings, or rubber mats provide cushioning and insulation against hard surfaces, promoting rest and relaxation for animals.

Ventilation

Proper ventilation is critical for maintaining air quality and controlling humidity levels inside shelters. Installing vents, windows, or fans helps remove stale air, moisture, and odors—preventing respiratory issues and heat buildup during warmer months.

Temperature Regulation

Implementing temperature-regulating features such as insulation, shade structures, and misting systems helps create a comfortable microclimate within shelters, reducing the risk of heat stress or hypothermia and promoting overall well-being.

Maintenance of Living Areas

Keeping the places where your animals live clean and germ-free is like giving them a safe haven where they can grow and thrive. Keeping these places clean isn't just a job; it's an important part of taking care of animals properly.

Maintenance Cleaning

To promote cleanliness and hygiene and stop the spread of diseases, living places must be maintained and cleaned on a regular basis. Getting rid of manure, replacing bedding, and disinfecting surfaces are all things that can be done to keep the surroundings healthy for animals.

Water Drainage

Making sure there is good drainage in and around shelters keeps water, mud, and stagnant puddles from building up, which can attract bugs and help diseases spread. Putting up gutters, French drains, or surfaces that slope away from living areas keeps the area dry and clean and reduces problems related to water.

Structural Integrity

Shelters are safe for animals and people when they are checked for signs of damage and wear, and addressed right away if any problems are found. Fixing roofs, walls, or floors that are damaged;

strengthening support beams; and changing worn-out parts can help keep people and animals safe and extend the life of shelter structures.

Chapter 10:

Marketing Strategies

(Dinnertime anticipation: the herd awaits the feed truck's arrival.)

As of 2018, the global demand for premium beef, such as cattle from rare breeds like the Belted Galloway, is expected to grow annually by 7.8% (Intelligence Hive, 2024). There has been a dramatic change in customer tastes towards premium meats that are sustainably farmed, and this increased demand is a reflection of that. More and more people are looking for alternatives to traditional beef breeds as a result of rising health and environmental consciousness among customers.

Belted Galloway cattle provide ranchers a chance to stand out from the competition and charge top dollar for their beef, which is in high demand throughout the world. Ranchers can take advantage of this rising demand and establish Belted Galloway beef as a top choice for customers who value sustainability and quality in their meat purchases by using marketing strategies that are specific to the breed. These strategies can emphasize the breed's history, health benefits, and exceptional flavor. Learn the ropes in this chapter!

Identifying Your Market Niche

If you want to be successful in the Belted Galloway cattle industry, or any industry, you need to know your market niche. Whether you're selling live animals or meat products, standing out in a crowded market is possible by identifying your niche and satisfying the demands and tastes of your ideal customers. Here, we'll go into the details of creating your niche in the Belted Galloway industry—including market dynamics, niche identification, customer preference, and trend analysis.

Understanding the Market for Belted Galloways

A small subset of beef consumers are drawn to Belted Galloways due to their unique look and high-quality flesh. Belted Galloways are highly regarded by both customers and chefs for their exceptional marbling, tenderness, and flavor—distinguishing them from standard beef breeds. Having said that, you should know that Belted Galloways, whether in meat or living form, have a small market when compared to other popular beef breeds.

Note: Selling Meat by the Half or Whole

One thing to keep in mind is that selling Belted Galloway as a whole is different from selling individual cuts. The Belted Galloway is usually sold as halves or full carcasses. By using this method, waste is minimized and ranchers are spared from having to deal with tough-to-sell particular portions of meat.

Finding a Niche

Recognizing a specific market need is crucial for ranchers who raise Belted Galloway cattle. Doing so allows them to tap into unrealized potential and stand out from the competition.

Organic Production

Belted Galloway cattle can find a niche in the organic cattle market. Belted Galloways raised utilizing organic ranching practices can appeal to health-minded and environmentally conscientious consumers, who are driving the demand for sustainably produced and organic food. Organic Belted Galloway beef can fetch premium rates, and ranchers can access this profitable market niche by getting certified and following stringent organic criteria.

Local Market

Providing services to specific geographic areas is another promising niche market possibility. There has been a recent uptick in the demand for farm-fresh and regionally produced foods. Ranchers can connect with consumers who appreciate honesty and transparency in their food supply by selling Belted Galloway beef directly to them through community-supported agriculture (CSA) programs, farm-to-table restaurants, and ranchers' markets.

Specialty Breeds

The beef market is ripe with opportunities for specialty breeds like the Belted Galloways. Compared to other types of beef, they stand out due to their distinctive look, heritage breed status, and high-quality meat. As a niche product, Belted Galloway beef allows ranchers to reach customers who value the breed's heritage, unique taste, and environmentally friendly farming methods. By targeting a certain subset of consumers who value uniqueness and authenticity, ranchers can make a profit by setting their goods apart from the competition.

Analyzing Market Trends and Consumer Preferences

Ranchers that raise Belted Galloway cattle would do well to keep abreast of changing customer tastes, new market trends, and market conditions so that they may make educated judgments, adjust their approaches, and seize opportunities as they arise. Now, let's get right into it and take a look.

Wellness and Health

The growing interest in health and well-being among consumers is influencing their choices in beef. More and more, shoppers are looking for hormone- and antibiotic-free, leaner cuts of beef. These health-conscious tastes are well-suited to Belted Galloway beef, which is characterized by its natural flavor and lean marbling. Ranchers can attract customers looking for healthy, ethically produced meat by showcasing the breed's natural production methods and its positive effects on human health.

Sustainability

Sustainable practices are also having a significant impact on what people want to eat. More and more people are looking for products made from sustainable ranching practices because they care about the environment and the well-being of the animals. An eco-friendly alternative to conventional beef production, Belted Galloway cattle thrive in pasture-based systems and benefit from rotational grazing. Ranchers can appeal to customers who value sustainability in their buying decisions by highlighting the breed's dedication to sustainable agriculture and its small environmental impact.

Branding and Promotion

Promotion and branding are very important in the cutthroat beef production industry for three reasons: to influence customer opinion, increase brand loyalty, and boost sales. Ranchers that raise Belted Galloway cattle must connect with their target clients and stand out in the market by developing a strong brand identity and using successful marketing methods. We'll talk about the most important parts of branding and advertising for Belted Galloway cattle businesses in this part. These include creating a ranch brand and identity, devising effective marketing materials and strategies, and using digital marketing and social media to spread your message and reach more people.

Developing a Ranch Brand and Identity

Defining Your Unique Selling Proposition (USP)

The essence of any thriving brand is its appealing unique selling proposition (USP), which differentiates it from rivals. Establishing a distinct identity for your Belted Galloway cattle ranch and breed requires you to know and explain what sets your cattle apart. Your USP is the bedrock of your ranch brand and the magnet that draws in clients who connect with your principles and products—whether it's the breed's striking beauty, the high quality of the meat, or your dedication to sustainable ranching techniques.

Building the Elements of a Visual Brand

Important visual branding components in communicating the character and essence of your ranch to potential customers include logos, color palettes, and typefaces. What constitutes branding your Belted Galloway cattle enterprise visually? Think about using images that represent the breed's unique traits, including the belted coat pattern or the rough terrain the cattle live in. Consumers are more likely to recognize and trust your brand when its visual identity is consistent throughout all promotional materials.

Effective Marketing Materials and Strategies

Compelling Marketing Collateral

Investing in high-quality marketing assets—such as brochures, flyers, business cards, and signs—is crucial for successfully promoting your Belted Galloway cattle business. Your brand's message, products, and services can be conveyed and leave an impact on potential customers through these concrete touchpoints. Your marketing materials should be well-designed, aesthetically pleasing, and convey the special qualities of your ranch and breed of cattle.

Targeted Marketing Strategies

To get the most out of your marketing budget, you need to zero in on your ideal clientele. If you want to connect with your target audience on a deeper level and generate engagement and sales, customize your marketing methods to resonate with them. This is true whether you're catering to health-conscious consumers, environmentally aware foodies, posh eateries, or specialist markets. You can contact potential clients who recognize the value proposition of Belted Galloway beef by exploiting specialized marketing channels including farm-to-table events, specialty food fairs, and agricultural trade exhibitions.

Utilizing Digital Marketing and Social Media

Harnessing the Power of Digital Platforms

If you want to reach and engage with customers in the current day, you need a strong online presence. You may reach clients all over the world and promote your Belted Galloway cattle business with the use of digital marketing channels like websites, email newsletters, and social media platforms. To draw in customers and keep them coming back, it's a good idea to have a professionally created website that features your ranch, animals, and products.

Leveraging Social Media Platforms

You may promote your Belted Galloway cattle business, increase brand recognition, and communicate with your community through social media sites like Facebook, Instagram, TikTok, and Twitter (X). Post interesting videos and images of your cattle, insights into ranch life, and facts about the breed's perks and quirks. Responding to comments, holding Q&A sessions, and holding competitions or giveaways are all great methods to engage with your audience, encourage involvement, and build brand awareness.

Direct Marketing, Ranchers' Markets, Online Sales

Ranchers that raise Belted Galloway cattle have a one-of-a-kind chance to interact with consumers through direct marketing. They can skip the middlemen and get to know their customers on a more personal level. Direct marketing, online sales, and ranchers' markets are just a few of the sales methods that will be discussed in this part.

Pros and Cons of Different Sales Channels

Direct Marketing

Pros:

- Having one-on-one conversations with consumers paves the way for more customized service and stronger relationships.
- Profit margins are higher because ranchers get to keep all the money they make.
- It's an opportunity to inform buyers about the breed, way of life on the ranch, and the quality of the meat.

Cons:

- Handling sales, client inquiries, and logistics demands a substantial investment of time and energy.

- It's not as widespread as more conventional means of distribution.
- You could potentially be confronted with regionally specific regulatory and licensing hurdles.

Ranchers' Markets

Pros:

- You'll gain entry to a captive market of nearby residents looking for farm-fresh, hand-picked food.
- It's the perfect chance to highlight Belted Galloway beef and other locally made products.
- Establishing a personal connection with clients cultivates a sense of trust and loyalty.

Cons:

- Depending on the time of year, ranchers may have fewer sales chances in the market.
- There might be rivalry from other sellers peddling comparable goods.
- Transportation, equipment, and market fees all pose costs.

Online Sales

Pros:

- You'll gain massive exposure to buyers all around the world who are looking for specialized beef products.

- Costs are lower than those of traditional brick-and-mortar stores.
- You can attain mastery of digital marketing technologies for hyper-specific promotion and advertising.

Cons:

- Financing online payment systems, website development, and e-commerce infrastructure is necessary.
- There is competition from other web stores and big beef producers.
- Logistics and shipping may be a real pain, especially when dealing with products that have a short shelf life, like fresh beef.

Setting Up a Direct Sales System

Identify Your Target Market

First things first: Figure out who you're selling to and what they want before you build up a direct sales method. Is your target market more concerned with health, foodies, or high-end eateries? One way to better satisfy the wants and needs of your target audience is to modify your sales strategy accordingly.

Establish Clear Pricing and Policies

To gain your clients' trust and faith, you must have transparent pricing and clear policies. Make sure that all costs, including delivery and processing, are clearly communicated when ordering Belted Galloway meat. Your online store's terms and conditions, including your policy on returns and refunds, should be clearly stated.

Streamline Order Fulfillment and Logistics

To guarantee that your clients receive their Belted Galloway beef on time, it is essential that your order fulfillment and logistical processes are efficient. Make sure the meat stays fresh and delicious by using dependable packaging materials and delivery procedures. Think about teaming up with reliable shipping companies or neighborhood delivery services to cut down on delivery time.

Tips for Successful Online Marketing and Sales

Build an Easy-to-Navigate Website

Your website is the online shop window for your Belted Galloway cattle enterprise. Generate a website that is attractive to the eye, simple to use, and mobile-friendly by hiring a professional. To improve the buying experience and establish credibility, it is recommended to include high-quality photographs, thorough product descriptions, and customer testimonials.

Discounts and incentives can be a great way to draw in new consumers, keep old ones, and get them to buy from you again and again. To encourage customer loyalty and boost sales volume, think about implementing loyalty rewards programs, special promotions, or discounts for big orders. To get the most out of your marketing efforts, track how well your campaigns are doing and make adjustments as needed.

Chapter 11:

Cattle Shows and Competitions

(Hauling our crew: Our trusty 24-foot stock trailer gets the job done.)

Over 2.5 million people attend the Houston Livestock Show and Rodeo annually (Homer, 2024). Cattle shows and contests continue to be a major part of the agricultural sector, as seen by the incredible turnout. Even outside of the arena, cattle exhibitions and contests have a significant impact. The cattle operations rely on these gatherings as key platforms for learning, growth, and development. During the show, exhibitors and breeders talk shop, exchange tips and tricks, and get the lowdown on what's new in the world of cattle genetics and breeding.

Additionally, heritage breeds such as the Belted Galloway are greatly assisted by cattle shows and contests. The breeders' efforts to highlight the special traits of these animals help get the word out about the need to preserve them for future generations at the ranch.

Moreover, cities that host cattle exhibitions and competitions see an uptick in tourism, business, and investment as a result. The economic viability of the region is boosted by the influx of participants and visitors to these events, which is good for local companies like hotels, restaurants, retail shops, and entertainment venues.

Cattle shows and contests are an intriguing topic, so let's learn more about them and how they have shaped the beef business.

Preparing for Shows: The Path to Success in the Show Ring

Showing cattle is an art form that demands a great deal of preparation, including study of the show ring's norms and expectations, thorough planning, and focus on detail. Let's take a look at the fundamentals of getting ready for performances!

Choosing Show-Ready Cattle

The first and most important stage in getting ready for a cattle show is picking out the best animals to enter. Breeders need to take the time to assess their herd and find the animals who have all the qualities that the show is looking for. Breed standards for Belted Galloway cattle may specify desired characteristics including coat color, body type, and muscular definition.

Conditioning and Grooming

The next step for breeders is to prepare and groom the show cattle so that they can perform to their maximum potential if they are chosen. This can include taking the animals in for frequent grooming appointments to maintain clean and shining coats, cutting and styling their hair to highlight their features, and making sure they get enough exercise and healthy food to keep them in top physical shape.

Understanding Judging Criteria

Going into a cattle show is more than just putting the animals on display; it's also a chance to see how well they conform to breed standards, how well they present themselves, and how well they follow instructions. Cattle show competitors demand that breeders understand the nuances of judging criteria in order to choose how to best present their Belted Galloway cattle. Let's take a look at some.

Breed Standards

Judging cattle in a show ring requires an understanding of the specific traits and requirements of each breed. Breeders of Belted Galloway cattle should study up on the breed's unique standards, which cover things like acceptable coat color, markings, dimensions, and temperament. When breeders are aware of these characteristics, they can choose and train cattle that will do well at shows.

Showmanship

Judges look at the animals' physical characteristics, but they also consider the exhibitors' showmanship abilities. To be a good showman, you need to know how to confidently and expertly handle your Beltie so that it presents its best side to the judges. To guarantee a flawless and professional presentation in the ring, exhibitors must practice basic showmanship tactics.

Logistics of Attending Shows

There is more to getting ready for a cattle show than simply gathering your cattle. Getting to the event on time, staying overnight, and packing all of your necessary items are all part of the preparations. Let's take a look at the practical side of going to shows and how to make sure everything runs smoothly for both you and your Belted Galloway cattle.

Registration and Entry

Breeders can't compete in a cattle show without first registering their animals. To do this, you may need to submit an entry form, pay an entry fee, and provide paperwork such as health certificates and animal registrations.

Transportation

To guarantee the comfort and safety of the show cattle throughout the voyage, meticulous preparation and coordination are required for their transportation to and from the venue. Breeders need to make sure that their transportation vehicles have enough air, bedding, and safe walls to keep the animals as comfortable as possible and avoid any harm while they are in route.

Accommodations

For shows that last more than one day, breeders must find places to stay near the venue for themselves and their animals. For the humans involved, this might include reserving campsites or motels; for the cattle, it could mean finding a stable or stall. The key to a successful exhibition is providing clean, pleasant housing for the animals and people who will be attending.

To ensure the well-being of their cattle throughout the show, breeders should also bring the necessary equipment and supplies. A variety of equipment is needed to fulfill the special demands of the animals, such as grooming tools, feed and water buckets, halters and leads, first aid kits, and whatnot. If breeders bring all the supplies they need for their show kits, they can give their animals the attention they need all day long.

Tips for Success in Competitions

It takes more than well-groomed cattle to win at cattle shows; it also takes perfecting your presentation abilities, working on your showmanship, and always pushing yourself to do better.

Get a Head Start

Cattle shows require extensive pre-event preparation. Choose animals that will do well in the show and get them started on their conditioning and training program early. There will be plenty of time for the animals to get in shape for the ring, which includes building muscle, improving coat quality, and generally becoming show-ready.

Exercise Every Day

To improve your animals' strength, stamina, and agility, make exercise a regular part of their daily regimen. To prepare them for the surroundings and what to expect in the show ring, this may involve going for walks or jogs on a leash, working on their halter skills, and exposing them to various hazards.

Healthy Eating

For the sake of your animals' physical and mental well-being, it is important to feed them a varied and balanced diet. Belted Galloway cattle have unique dietary requirements depending on their age, weight, and degree of activity; a nutritionist or vet can help you determine these needs.

Practice Makes Perfect

Make it a priority to work on your animal presentation and showmanship abilities. Among these skills is the ability to stand and walk elegantly on a lead, as well as how to correctly prepare for an evaluation and respond to directions with composure and assurance in a show ring.

Attention to Detail

Your animals' temperament and performance in the ring are just as important as their grooming and training. Make sure their coats are clean and well-groomed, their hooves are trimmed, and that their entire look shows your commitment to excellence and the breed standard.

Confidence and Poise

Display your animals with self-assurance and grace in the show ring. Always stand up, look the judges in the eyes, and control your emotions when dealing with the animals. Giving off an air of self-assurance does wonders for your performance and for making a good impression on the audience and judges.

Learning from Competition Experiences

Ribbons and medals aren't the only things exhibitors can win at cattle shows; these events also provide great educational opportunities. Let's take a deep dive into the realm of competitive learning and find out how every show ring experience may lead to better success and expertise in the cattle sector.

Reflect and Evaluate

Think about how you did and where you can make changes after each tournament. Think about the judges' or other exhibitors' comments, your own presentation and showmanship abilities, and the animals' strengths and shortcomings.

Continuous Improvement

Embrace the chance to learn and grow from competition experiences. Learn more about breed standards and judging requirements, hone your presenting skills, and improve your training and conditioning routine.

Networking and Collaboration

Participate in cattle exhibitions to network with other farmers or ranchers, breeders, and industry experts; this is a great place to pick their brains and learn from their experiences. Opportunities for mutual success, support, and new perspectives can arise through networking and teamwork.

Conclusion

Preparation, commitment, and a thirst for progress are the three pillars upon which success in cattle shows rests. You can improve your performance as a Belted Galloway exhibitor and reach your goals by devoting time and energy to training and conditioning your animals, honing your presentation and showmanship abilities, and using competitions as chances to learn and grow. The path to greatness and the connections made along the way are more important than any number of ribbons you may win.

Chapter 12:

Routine Ranch Tasks

(Fresh start: A few hours old and soaking up the sunshine.)

Ranch life is not possible without the regular chores that keep the animals and the land healthy and safe. These mundane but essential tasks ensure the ranch runs smoothly and productively, even though they aren't often fun or glamorous.

Maintaining the animals' health and well-being is the primary goal of everyday tasks. A daily routine that includes feeding, watering, and monitoring livestock is essential for their health and well-being. By keeping tabs on their livestock and checking for any signs of illness, ranchers can ensure that their herds are healthy and happy.

Furthermore, regular chores contribute to the upkeep of the ranch's infrastructure. These routine maintenance tasks, which range from fixing fences to repairing equipment, keep little issues from becoming big hassles. Ranchers maintain their land and livestock for future generations by keeping up with routine maintenance and repairs.

Beyond that, the ranch as a whole benefits from the efficiency and production brought about by routine duties. Ranchers can cut down on wasted time and effort by sticking to a regular schedule for housekeeping tasks. Not only does this save money, but it also frees up ranchers' time to concentrate on other crucial areas of business like advertising, strategy, and creativity.

Come ride along as we unpack the traditions and responsibilities that make up ranch life, one chore at a time, and see how they entwine to form the complex world of husbandry. Here at the ranch's beating heart, you can find adventure and the reward of a job well done in every task.

Daily, Weekly, and Monthly Checklists

For ranchers, checklists—whether daily, weekly, or monthly—are a lifesaver because they lay out a plan for handling all the many moving parts that make the business tick. Make sure you don't miss a single important task by using these checklists for everything from watering and feeding livestock to fixing machinery and organizing supplies. Let's explore why these checklists are so useful and how to put them to good use on the ranch.

Daily Checklist

The daily checklist is the lifeblood of ranch operations; it includes all the things that need to be done every day to keep the infrastructure and cattle healthy. In most cases, this checklist will contain items like:

- **Preparing food and water for animals:** The well-being and efficiency of animals depend on their access to sufficient food and water.

- **Medical examinations:** Ranchers can reduce the likelihood of disease outbreaks and unnecessary suffering by keeping a close eye on their cattle on a regular basis for symptoms of disease or damage.

- **Inspection of the fence and facility:** Looking for damage or wear on fences, gates, and other structures helps keep the animals safe and secure and keeps them from getting out.

- **Urgent or time-sensitive tasks:** To keep minor issues from becoming major disasters, it is critical to handle any urgent matters or emergencies as soon as they arise.

Weekly Checklist

For the sake of the ranch's efficiency and usefulness, the weekly checklist dives into things that need doing less often but are nonetheless crucial. This checklist may include:

- **Maintenance of fences:** Fixing damaged or broken fences aids in containing livestock.

- **Equipment servicing:** Making sure that machinery and equipment are regularly maintained and serviced helps to avoid breakdowns and downtime, which in turn keeps operations running smoothly.

- **Supply replenishment:** Maintaining a good supply of things like feed, bedding, and medical tools for the animals makes sure the ranch has everything it needs to run smoothly.

- **Maintenance and clean-up tasks:** Doing bigger jobs like cleaning the barn, mowing the field, or organizing the equipment shed keeps the workplace clean and functional.

Monthly Checklist

While certain items on the monthly checklist may not be as pressing as others, their importance to the ranch's well-being, productivity, and longevity cannot be overstated. This checklist may include:

- **Analyzing and testing soil:** To maximize production and sustainability, ranchers can optimize pasture management and fertilization by assessing soil health and nutrient levels.

- **Organizing the rotation of pastures:** Livestock pasture rotation helps keep soil fertility high, reduces parasite infestations, and stops overgrazing.

- **Fiscal audit:** Ranchers can keep tabs on spending, find places to cut costs or make investments, and check the operation's financial health by reviewing financial records and budgets.

- **Visiting the vet and getting vaccinated:** Making sure that animals get regular medical visits and vaccinations helps stop disease outbreaks and supports their health and well-being.

Ranchers can improve their organization, resource allocation, and task prioritization by dividing chores into daily, weekly, and monthly categories. These checklists are a great way to stay organized and on top of important tasks, and ensure the ranch is successful and sustainable.

Seasonal Adjustments in Care

Let's take a look at why it's important to tailor your care to the changing seasons and how you may optimize your livestock management methods to conquer the challenges of each. This list depends on where you live and your weather.

Summer

Heat stress, thirst, and a lack of food are all potential problems for cattle in the summer. Modifications to care plans can be made based on the season. This can involve adding cooling systems and shade, making more water available, and altering feeding schedules to account for changes in fodder availability and quality.

Winter

The primary goals of winter livestock management are to keep animals warm and safe from the severe weather and nutritional difficulties caused by the scarcity of fodder. To keep livestock healthy and comfortable throughout the winter, it is important to provide them with shelter, additional food, and water. This will help prevent cold stress. Install water heaters or make use of solar power if you are not near an electrical outlet. Warm water is beneficial for animals, including cattle, during the winter months. This keeps them from getting cold on the inside, which keeps them at a more normal internal temperature.

Spring and Fall

During the spring and fall, there is a greater chance of disease transmission, variations in the quality of the feed, and temperature fluctuations. Depending on the time of year, it may be necessary to make adjustments to the herd's care in order to prevent parasites, manage reproduction, and improve nutritional intake through forage and grazing.

In order to keep cattle healthy, happy, and productive all year round, it is vital to use consistent care procedures. Regular care on a daily, weekly, and monthly basis allows ranchers to see problems early, treat them effectively, reduce the likelihood of stress and illness, and improve health, in general. Animals benefit from predictability and stability, which decreases anxiety and promotes a sense of security in the herd, when care techniques are consistent.

Prioritizing and Scheduling Tasks

An essential part of managing a ranch is making a schedule and prioritizing work so that ranchers may use their time and resources wisely. Ranchers can keep their animals and land healthy, produce more, and experience fewer interruptions if they prioritize their work and plan it accordingly.

Task Prioritization

Before you can begin to prioritize your projects, you must determine which ones are the most critical or establish a deadline. That means thinking about things like how it would affect animals' well-being, how much money could be lost, and what might happen in the future. The health of the animals and the ranch's bottom line are directly impacted by chores like feeding, watering, and checking on their vital signs; therefore, these tend to be high on the list of priorities. Also, less important things take a back seat to more pressing matters, like replacing a damaged fence or tending to a sick animal.

Distributing Assets

Ranchers need to prioritize their work and then divide up their time, labor, and tools appropriately. Finding the best approach to allocate available resources to tackle high-priority projects is part of this process. For instance, in the event that a storm is anticipated to strike the region, it is imperative that ranchers take necessary measures to secure any loose equipment and strengthen shelters. They may have to reallocate workers and equipment, as necessary, to accomplish these activities in time for the storm to strike.

Timely Task Scheduling

The key to efficient task scheduling is taking into account variables like workload, weather, and available resources to establish the best time to do high-priority tasks. To avoid breakdowns and downtime during peak operational periods, it is recommended to schedule routine maintenance and equipment servicing during periods of low activity. The same holds true for soil testing and pasture rotation; they are best performed at certain periods of the year to take advantage of favorable conditions and maximize sustainability.

Flexibility

Efficient ranch management requires both the ability to plan ahead and prioritize activities, along with the agility to respond to unexpected challenges. Ranchers need to be flexible and ready to change their plans on the fly in case of crises or unanticipated situations. In light of fresh data or altered circumstances, this may necessitate rearranging existing priorities, redistributing resources, or delaying some tasks.

Collaboration and Communication

The ranching crew needs to be able to talk to each other and work together if they want to prioritize and schedule activities well. For ranchers, the key to a well-coordinated effort is open and frequent communication of objectives, timelines, and expectations. The ranching operation promotes a culture of responsibility and teamwork through frequent meetings and check-ins that provide for the sharing of updates, comments, and necessary plan revisions.

Delegating and Managing Staff Responsibilities

Managing staff roles and distributing work becomes crucial for larger livestock operations, during hectic times, to stay productive and efficient. Regular checklists—whether they're daily, weekly, or monthly—may be a great way for ranchers to stay in touch and make sure that everyone is on the same page when it comes to delegating work.

Ranchers can ensure that daily operations, like feeding and checking on cattle, are done quickly and reliably by assigning these responsibilities to specific individuals or teams. The same holds true for jobs that need specific knowledge or abilities, like maintaining machinery or keeping accurate financial records; delegating these responsibilities to competent employees allows you to play to their strengths and guarantees top-notch results.

Another way ranchers may encourage responsibility and cooperation is by making sure everyone has a specific task to do every day, week, and month. Adjustments can be made as needed through regular communication and feedback loops, which ensures activities are executed properly and any difficulties or concerns are addressed swiftly.

Last but not least, ranchers can't do their jobs well without checklists—daily, weekly, and monthly—that lay out the many moving parts necessary to keep the ranch running smoothly. Ranchers may make sure their operation operates well, profitably, and sustainably in the long run by making detailed checklists, correctly scheduling and prioritizing chores, and assigning duties to employees, as needed.

Maintenance Schedules for Equipment and Facilities

Regular equipment and facility maintenance is an important part of ranch management since it keeps the tools and infrastructure needed for everyday activities running smoothly. For ranchers, the key to maximizing the efficiency, safety, and lifespan of their facilities and equipment is to create regular maintenance routines, keep thorough records of repairs and updates, and plan for renovations and replacements as needed.

Regular Maintenance Routines

Preventing breakdowns, increasing equipment lifespan, and keeping ranch facilities in good repair can all be achieved through the establishment of regular maintenance routines. This entails checking and servicing on a regular basis according to a schedule, fixing problems before they become dangerous or expensive to fix. Regular equipment maintenance may involve changing the oil, replacing the filters, lubricating, and calibrating the device. Some examples of what can fall under the umbrella of "facilities maintenance" are cleaning, painting, fixing structural damage, and inspecting for indicators of wear and tear. Ranchers can limit downtime, save money on repairs, and keep their facilities and equipment in top shape if they plan ahead for maintenance.

Keeping a Log of Repairs and Updates

It is crucial to keep meticulous records of all updates and repairs to trace the history of facilities and equipment, find repeated problems, and make educated judgments regarding replacements and maintenance. Ranchers should maintain a record that details each repair or update—including the date, the problem or improvement's nature, the materials or parts utilized, and the name of the person responsible for the work. This log is a great resource for troubleshooting, planning maintenance, and documenting the overall health and performance of facilities and equipment over time. It may also be required to keep records of repairs and updates for warranty claims, insurance, or regulatory compliance.

Planning for Upgrades and Replacements

The ranch's infrastructure and tools may become inefficient, unreliable, or unfit for purpose as they get older. To keep the ranch competitive, safe, and productive—it's important to plan for updates and replacements. Regularly evaluating the state and functionality of facilities and equipment, pinpointing areas that require repair or replacement, and formulating a strategy for their execution are all part of this process. Technological developments, shifts in operational requirements, cost-benefit analysis, and sustainability in the long run are all factors to think about while organizing replacements and upgrades. Ranchers may stay ahead of the competition, increase efficiency, and save money by planning ahead for renovations and replacements.

Ranchers may minimize downtime, repair costs, and safety hazards while maximizing the lifespan and performance of their equipment and buildings by establishing regular maintenance routines, keeping thorough records of repairs and updates, and planning for upgrades and replacements, as needed. The ranch's efficiency, production, and profitability are all enhanced when time and money are invested into maintenance.

Chapter 13:

Recordkeeping

(With my double-layer of hair, I'm not cold.)

Recordkeeping is an important part of Belted Galloway cattle ranching. It's the basis for good management, making decisions, and long-term success. Extensive recordkeeping helps ranchers with various tasks like keeping tabs on the well-being of individual animals, keeping an eye on the results of breeding, handling financial transactions, and making sure they're in compliance with regulations. This all adds up to valuable information that ranchers can use to make their operations more productive, profitable, and sustainable.

The next part will take you on a journey into the recordkeeping—unveiling the methods, resources, and standards that support this vital task. Come with us as we explore the ranch's core, where careful paperwork turns facts into knowledge and figures into insights.

Importance of Detailed Records for Belted Galloway Cattle Ranching

If you want to ranch Belted Galloway cattle successfully, you need to keep good records. If you keep detailed records of your business dealings, from the medical history of each animal to the money you make and lose, you may learn a lot about the well-being, output, and profitability of your herd.

Tracking Cattle History, Health, and Productivity

Individual Animal Records

You may monitor the progress of your cattle herd's history, health, and output by keeping meticulous records for each individual animal. Data like weights and growth rates—as well as information like dates of birth and parentage, immunizations, medical treatments, and breeding records—are all part of this. With this data at your fingertips, you can maximize your herd's genetic potential through well-informed decisions on breeding plans, health management, and culling.

Herd Health Management

If you want to keep an eye on and manage the health of your Belted Galloway herd, you need to keep records. You can safeguard your animals' health and wellbeing by keeping track of their immunization schedules, deworming treatments, illness outbreaks, and trips to the vet. This will allow you to spot trends, identify new health problems, and take preventative actions. Complying with certification programs and regulatory standards, such as those for organic or grass-fed ranching, also relies heavily on detailed health records.

Performance Tracking

Records are very helpful for figuring out how productive and well-behaved each animal is and how well the whole group is doing. You can find the best animals, spot trends, and make choices based on data to increase the productivity and profitability of your herd by keeping track of things like calving rates, weaning weights, feed conversion ratios, and reproductive efficiency. You can compare your operation's performance to industry standards and find areas for improvement with the benchmarking data provided by performance records.

Legal and Financial Recordkeeping

Regulatory Compliance

To stay in line with all the rules and regulations that govern cattle ranching, it is crucial to keep meticulous records. Environmental rules, food safety requirements, and animal welfare recommendations must be followed, and records of animals' identifying information, whereabouts, health, and treatments must be kept. To avoid fines and other consequences for noncompliance, it is important to have records that are accurate and up-to-date. This will help prove compliance throughout certification, inspection, and audit processes.

Financial Management

When it comes to Belted Galloway cattle ranching, recordkeeping is key for efficient financial management and planning. You can learn a lot about the financial health and success of your business by keeping tabs on all of the money coming in and going out. This includes the money spent on feed and veterinarian care, as well as on equipment, labor, and marketing. If you keep meticulous financial records, you can find out where money is going, how to best put it to use, and how to set prices, make investments, and divide up resources so that you can be profitable and sustainable.

Using Records for Decision-Making and Planning

Informed Decision-Making

People who raise Belted Galloway cattle can use detailed records to help them make decisions. You can find trends, patterns, and ways to make things better by looking at past data and performance measures. This will help you make smart choices about breeding, managing health, marketing, and allocating resources. You may use records to make evidence-based decisions that drive your operation's success—whether it's choosing breeding stock, changing feeding programs, or implementing new management techniques.

Strategic Planning

You can't plan in advance or envision the future of your Belted Galloway cattle ranch without records. Planning ahead—whether it's to diversify your product offerings, increase your herd size, or break into new markets—is made easier with the help of market trends, industry benchmarks, and performance statistics from the past. If you want to be competitive and adaptable in the ever-changing agricultural scene, keeping detailed records is essential for tracking your progress toward goals, measuring your success, and adapting your methods, as needed.

Tools and Systems for Effective Recordkeeping

Nowadays, having good records is important for managing herds better, following the rules, and making the most money from Belted Galloway cattle ranching. Modern technology has provided ranchers with an abundance of resources that can improve operating efficiency and simplify recordkeeping.

Traditional vs. Digital Recordkeeping Systems

In the past, ranchers would keep track of important herd data using pen and paper or simple spreadsheet programs. Although these approaches are acceptable for smaller operations, they aren't always up to the task of managing bigger herds or more complicated data sets. There is a risk of inaccuracy and inefficiency in herd management due to the time and effort required for human recordkeeping.

When compared to more conventional approaches, digital recordkeeping systems provide ranchers a number of benefits, such as the ability to easily automate computations, provide thorough reports, and simplify data entry. Digital systems can be as basic as pre-made spreadsheets or as complex as ranch management software developed with cattle ranchers in mind. Ranchers may increase data accuracy and operational efficiency, and obtain vital insights into herds' performance and productivity, by embracing digital technologies.

Ranch Management Software and Applications

CattleWorks

Developed with the needs of cattle ranching operations in mind, CattleWorks is an all-inclusive computer program. With this robust program, you can easily manage your herd in every way: from keeping records to managing breeding, monitoring health, doing financial analyses, and much more. Ranchers can use CattleWorks to enter and keep track of detailed information about each animal, make reports that are tailored to their needs, and make smart choices that will improve the performance and profitability of their herd.

Other Ranch Management Software

Other than CattleWorks, there is a range of ranch management software to choose from, each with its own set of advantages and disadvantages. A few well-known instances are:

- **CattleMax:** This is a cloud-based ranch management software that lets ranchers keep cattle records, manage breeding and health data, and analyze performance indicators from any internet-connected device.

- **Cow Sense:** It's a herd management software for cow recordkeeping, financial analysis, genetic evaluation, and reproduction.

Data Security and Backup Strategies

Data Security

Belted Galloway cattle ranchers place a premium on safeguarding confidential ranch records from prying eyes. Protecting sensitive information and avoiding data breaches is easier with strong data security mechanisms in place, such as encryption, passwords, and user access controls. Keep your systems safe from ever-changing cyber threats by updating and auditing them regularly.

Backup Strategies

Cattle ranching operations are extremely vulnerable to data loss; hence, it is crucial to have a solid backup plan in place to safeguard important ranch records. Protect your data from hardware failure, natural calamities, and human error by backing it up regularly to a secure off-site location. This might be an external hard drive, cloud storage, or a storage service. Consistent and easy access to mission-critical data in times of crisis is guaranteed by automated backup solutions.

Chapter 14:

Ranch Safety

(Sunrise on the Rebel Ranch: A new day dawns on the ranch.)

Cattle ranching poses its own distinct safety concerns, but did you realize that agriculture is one of the most dangerous industries in the world? Unpredictable cattle behavior and operating heavy machinery are only two examples of the inherent risks in ranching that necessitate constant adherence to safety rules and practices. To make sure that ranchers and Belted Galloway cattle are both safe, let's look at why ranch safety is so important and how to make it less dangerous.

Safety Protocols and Training

Ranching Belted Galloway cattle requires a high level of safety awareness due to the numerous dangers that can befall people and livestock in this ever-changing landscape. To reduce hazards and guarantee a safe working environment, this section will discuss the significance of creating thorough safety procedures and giving thorough training to ranch employees. Discover the fundamentals of ranch safety training and procedures, including creating a safety plan and holding frequent safety drills and updates.

Designing a Ranch Security Plan

To begin developing a ranch safety plan, it is necessary to undertake a comprehensive risk assessment in order to identify any risks and weaknesses in the operation. Possible examples of such dangers include working with animals, operating heavy machinery, dealing with potentially dangerous substances, and various environmental variables. Ranchers can improve overall safety by identifying potential dangers and developing focused methods to reduce or eliminate them.

Following the identification of potential dangers, it is critical to establish detailed protocols for safe work practices. Make sure that all ranch staff can quickly access these procedures and that they are clear and succinct. Among the many topics they should address are fire safety, emergency response procedures, chemical handling, machine operation, and livestock handling.

Along with taking precautions, ranchers should have a detailed strategy for handling any kind of emergency, including accidents, injuries, and other similar situations. Procedures for contacting rescue services, administering first aid, evacuating employees, and collaborating with local authorities should all be part of this strategy. The efficiency of the emergency response plan in times of crisis is ensured by regular assessment and modifications.

Training Staff in Safety Procedures

A comprehensive safety training session is necessary for every member of the ranch crew—whether they are paid employees, independent contractors, or volunteers. Methods for safely handling cattle, operating machinery, preventing fires, and responding to emergencies are all part of what this training should include. To ensure that new employees are well-versed in ranch-specific safety protocols, it is imperative that they complete full orientation and receive ongoing on-the-job training.

Safety training isn't a one-time thing; it's an ongoing process that needs to be continued and improved on a regular basis. Regular refresher training sessions should be conducted by ranchers to review safety practices, go over any new equipment or processes, and address any new safety problems that may arise. Staff members are kept alert, educated, and ready to deal with possible dangers via continuous training.

Regular Safety Drills and Updates

If you want to make sure your staff is ready to handle crises and that your safety procedures are working, you need to hold frequent drills. In these drills, employees can practice responding to a variety of situations in a safe setting, such as when livestock escapes, when machinery breaks down, or when chemicals are spilled. Participation in post-drill debriefing meetings allows for the exchange of ideas and suggestions for enhancement.

It is essential to revise safety procedures when ranch activities change and new dangers appear. To keep up with the latest developments in machinery, software, rules, and industry standards—ranchers should assess and update their safety protocols on a regular basis. It is important to keep safety protocols up-to-date so that they can effectively address the dangers and problems that the organization is now encountering.

Emergency Preparedness

Belted Galloway cattle and ranchers alike are vulnerable to the unpredictability of emergencies and natural catastrophes. To mitigate hazards, protect lives, and defend ranch operations, emergency preparedness is vital, as we will explore in this section.

Emergency Preparedness in the Face of Natural Disasters

A thorough risk assessment should be carried out to identify possible dangers and weaknesses unique to the ranch's setting and surroundings as the initial stage in disaster preparation. Wildfires, floods, storms, and earthquakes are all examples of natural disasters that need risk assessments. Human-caused emergencies, including chemical spills or mechanical mishaps, also need risk assessments. Ranchers can lessen the likelihood of adverse events and their effect on operations by identifying and assessing potential risks.

A thorough emergency response plan, detailing methods for reacting to different types of emergencies, should be developed by ranchers based on the results of the risk assessment. Included in this strategy should be procedures for alerting staff, removing animals from the area, protecting property, and working in tandem with rescue organizations and local governments. Staff members are more prepared to handle emergencies when they participate in regular drills and training exercises.

Emergency Kits and Supplies

Animal Care Supplies

In case of an emergency, have a supply kit ready to go with everything you need to care for Belted Galloway cattle. A list of what should be in an emergency pack is as follows:

1. **Feed:** Include a supply of feed appropriate for the dietary needs of Belted Galloway cattle, such as hay or grain.

2. **Water:** Provide a sufficient quantity of clean water to ensure hydration for the animals.

3. **Medications:** Include any necessary medications or supplements prescribed for the cattle, such as antibiotics or vitamins.

4. **First aid supplies:** Stock the kit with basic first aid supplies for treating minor injuries or ailments—including bandages, antiseptic solution, and wound dressings.

5. **Halters and lead ropes:** Have halters and lead ropes on hand for safely handling and restraining the cattle during emergencies.

6. **Identification tags:** Ensure that each animal has identification tags with important information such as name, owner contact information, and any medical conditions or special needs.

7. **Blankets or tarps:** Provide blankets or tarps to create temporary shelter or protection from inclement weather conditions, such as rain or cold temperatures.

Make sure the emergency kit is always packed and ready to go by checking and restocking its contents on a regular basis, especially feed and water. Put together an emergency kit and keep it somewhere you can readily find it in case of an emergency.

Personal Safety Gear

Ranchers shouldn't only stock up on supplies for their animals; they should also put together emergency kits for themselves and their employees, along with vital safety equipment. Everyone working on the ranch should have an emergency kit that is specialized for their requirements and prepared for the kinds of crises they might face. In the event of an emergency, a rancher's personal safety equipment should include the following:

1. **protective clothing**

2. **sturdy footwear**

3. **gloves**

4. **helmets**

5. **flashlights**

6. **batteries**

7. **multi-tools**

8. **personal hygiene items**

Communication and Evacuation Plans

Get the ranch staff, nearby properties, and the necessary authorities on the same page regarding how to communicate important information in the event of an emergency. Some examples of such channels are community alert systems, social media, text messaging, mobile phones, and two-way radios. In times of crisis, assign particular people or groups the task of coordinating communication efforts.

Before an emergency occurs, make sure everyone knows how to get out of the area and where to meet up for additional instructions. These pathways should take people and animals safely out of danger zones and to prearranged meeting places, in the event of an evacuation or disaster. Review and update evacuation routes as appropriate, and make sure they are well-marked and recognizable to all workers.

Last but not least, being ready for any kind of catastrophe is essential for managing Belted Galloway cattle responsibly. This will protect everyone involved in the event of a disaster. Ranchers can safeguard their operations from potential danger by preparing for emergencies and natural catastrophes, stocking up on emergency supplies, and creating strategies for communication and evacuation. Success in the unpredictable world of cattle ranching requires a culture of safety and preparation, which may be fostered through an aggressive commitment to emergency planning. This method also increases resilience and readiness.

Conclusion:

Embracing the Future of Belted Galloway Cattle Ranching

(Our fur buddies looking out the door)

As we come to the end of our journey through the world of Belted Galloway cattle ranching, it's essential to reflect on the vital knowledge and best practices that have emerged. We have covered the basics of ranch management, including learning the breed's features, creating good breeding programs, and keeping everyone on the ranch safe. Yet, this is only the first step in what will undoubtedly be a long and winding road as we explore the world of Belted Galloway cow ranching.

Within this book, we have explored every aspect of Belted Galloway cattle ranching, revealing the keys to success in this age-old business. Breeding plans and genetic selection are crucial for producing high-quality cattle with desirable features, as we have learned.

We have looked at how herd health measures, pasture management, and appropriate nutrition can keep our herds healthy and productive. We have covered the importance of understanding proper livestock handling procedures, following safety rules, and being prepared for emergencies so that ranchers and cattle are both protected. All the while, we have stressed the importance of ranching in a way that is both environmentally and socially responsible, with a focus on the animals, the land, and the people we serve.

Future Outlook for Belted Galloway Cattle Ranching

There is a lot of hope and opportunity for the future of Belted Galloway cattle ranching. Although Belted Galloways have long been popular in the East, the breed is now gaining popularity in the West as ranchers discover the many advantages of this versatile animal. Because of their attractive look, high meat quality, and ability to adapt to different conditions, Belted Galloway cattle are expected to experience a boom in demand as the breed moves westward. Through hard work and creativity, we can ensure that Belted Galloway cow ranching remains a pillar of environmentally responsible farming for future generations.

It is our deep-seated duty as landowners and animal caregivers to maintain ranching techniques that are both environmentally responsible and morally sound. On our ranches, we should work to increase biodiversity, reduce our impact on the environment, and preserve natural resources. Taking excellent care of our cattle and making sure they have a good life should be our top priorities. And we need to be active members of our communities, connecting with people and creating relationships based on honesty, openness, and respect for one another. The future of Belted Galloway cattle ranching is bright and promising if we accept these principles and strive for constant progress.

Finally, keep in mind that raising Belted Galloway cattle is more than just a job; it's a way of life that has been preserved through the years, based on great respect for the land and the animals who dwell there. In the next leg of your trip, may we approach it with enthusiasm, optimism, and a relentless pursuit of greatness. Together, we can write the next great chapter in the story of Belted Galloway cattle ranching—one defined by innovation, sustainability, and enduring success.

We hope you enjoyed this book! Did this Belted Galloway content moo-ve you? Your thoughts and opinions are incredibly valuable to us. Sharing a star rating and a quick review on the website where you purchased it is like giving us a virtual *high five* (or hoof bump)! Your feedback helps others discover this book and allows us to keep creating content you'll love—it makes us do a happy cow dance!

Appendix:

Tips for Cooking Your Grass-Fed Beef

Grass-fed beef requires different cooking techniques. You must adjust your cooking times and temperatures if your used to purchasing your meat from a grocery store (normally grain-fed). Grass-fed beef is denser because the animals are leaner and roam freely on pastures. This does NOT make it tough; quite the contrary. The Belted Galloway (Beltie) breed has been tested to be one of the most tender and flavorful meat breeds available. However, ALL grass-fed beef must be cooked properly.

The main reason grass-fed beef (or any meat) becomes tough, is overcooking! Because of the denser meat, grass-fed beef usually requires 30% to 40% less cooking time and continues to cook much longer when removed from heat compared to grain-fed beef. Grass-fed beef is best cooked rare to medium-rare.

Two things will help you prepare the best steaks: a needle tenderizer and a meat thermometer. Before cooking, use a needle tenderizer to inject your liquid seasoning of choice. Garlic pepper and seasoned salt sprinkled on the outside work well, as do other BBQ spices. Use a meat thermometer and check it frequently when cooking. For rare steaks, I take mine off the grill when they are about 110 degrees. Then, I let the meat rest for at least 15 minutes.

If you like well-done beef, then cook grass-fed beef at very low temperatures, and you may want to add a sauce to add moisture, so it doesn't dry out. When grilling, sear the meat quickly over high heat on each side to seal in its natural juices and then reduce the heat to medium or low to finish the cooking process. Since grass-fed beef is extremely low in fat, lightly coat it with virgin olive oil, truffle oil, or a favorite light oil for flavor enhancement and easy browning when grilling. The oil will also prevent drying and sticking.

Stove-top cooking is excellent for any steak, including grass-fed steak! You have more control over the temperature than on the grill. You can use butter in the final minutes when the heat is low to carry the taste of fresh garlic through the meat as steak chefs do. In addition to smoking/cooking on the BBQ, we sometimes cook steaks in a cast iron skillet with bacon fat, or butter. I start at high heat to sear the steak, then reduce my temperature to medium and take the steaks out of the pan at 110 degrees. You can also cook onions and mushrooms in the pan with the steak. Both smoking/BBQ and pan cooking result in a tender, flavorful steak.

If roasting, reduce the temperature of your previous beef recipes by 50 degrees. This usually means around 275 degrees for roasting, or at the lowest heat setting in a crock pot. The cooking time will still be the same or slightly shorter, even at the lower temperature. Again, watch your meat thermometer and don't overcook your meat. Use moisture from sauces to add to the tenderness and prevent drying out when cooking your roast.

Also, baste to add moisture throughout the BBQ grilling process. Don't forget that grass-fed beef requires 30% less cooking time, so don't leave your steaks unattended. They will cook VERY fast compared to grain-fed. Use low temperatures on the grill so that you do not overcook. Smoke at a low temperature for an hour to add flavor, then turn up the temp to 225 degrees and cook to how you like it, making sure to check the temperature often with a thermometer.

When roasting, sear the beef to lock in the juices and place it in a preheated oven. Save your leftovers. Roasted grass-fed beef slices make good, healthy luncheon meats instead of “processed lunch meats." Slice a warm or cold roast and add fresh red onions, green peppers, provolone cheese and mayo on a toasted sub.

I truly hope you enjoy cooking and eating grass-fed beef. If you're used to cooking conventional store-bought meat, it may take some time to get used to the change associated with cooking leaner meats, but I think you'll find it well worth it.

Rebel Ranch

Glossary

1. **Artificial Insemination:** A reproductive technique in which semen from a bull is collected and manually introduced into the reproductive tract of a female Belted Galloway cow to achieve pregnancy.

2. **Belted Galloway Cattle:** A breed of beef cattle originating from the Galloway region of Scotland, known for their distinctive white belt around their midsection.

3. **Body Condition Score (BCS):** A method used to assess the health of livestock by evaluating their fat reserves.

4. **Branding:** The practice of marking Belted Galloway cattle with a unique symbol or design to signify ownership or identification.

5. **Breeding:** The process of mating Belted Galloway cattle to produce offspring with desired traits.

6. **Breeding Program:** A planned approach to mating Belted Galloway cattle with the goal of improving specific traits or characteristics in the offspring.

7. **Bull:** An adult male bovine.

8. **Calving:** The process of giving birth to a calf.

9. **Cattle Auction:** A public sale where Belted Galloway cattle are bought and sold for breeding or meat production purposes.

10. **Cow:** Female that has given birth.

11. **Cud:** Partially digested food that is regurgitated from the rumen of Belted Galloway cattle and chewed again for further digestion.

12. **Embryo Transfer:** A reproductive technique in which embryos from genetically superior Belted Galloway cows are collected and transferred into recipient cows to produce offspring.

13. **Forage:** Vegetation, such as grasses and legumes, consumed by Belted Galloway cattle as food.

14. **Forb:** Sometimes referred to as herbs, forbs are broadleaf plants that are not woody or grass-like. Rangeland forbs are often perennial and could possibly be poisonous, but don't usually overpower a pasture unless there has been serious overgrazing.

15. **Genetics:** The study of heredity and variation in living organisms, including the transmission of traits from one generation to the next in Belted Galloway cattle.

16. **Grazing:** The act of feeding on grass or other vegetation in a pasture or range, a primary feeding behavior of Belted Galloway cattle.

17. **Herd:** A group of Belted Galloway cattle kept and managed together on a ranch or farm.

18. **Heifer:** Female that has not given birth.

19. **Livestock Handling:** The skillful and humane management of Belted Galloway cattle during activities such as herding, loading, and transporting.

20. **NRCS:** The Natural Resources Conservation Services. It's a part of the US Department of Agriculture (USDA).

21. **Pasture:** A fenced-in area of land where Belted Galloway cattle are allowed to graze and forage for food.

22. **Personal Safety Gear:** Essential safety gear and supplies for ranchers and their staff, including protective clothing, sturdy footwear, gloves, helmets, flashlights, batteries, multi-tools, and personal hygiene items.

23. **Point of Balance:** Refers to a specific area on the animal's body that relates to its flight zone (the area it feels comfortable in) and its vision.

24. **Ranch:** A large farm or agricultural operation where Belted Galloway cattle are raised for meat production.

25. **Rancher:** A person who owns or manages a ranch and is responsible for the care and management of Belted Galloway cattle.

26. **Riparian Buffer:** Also known as a "stream buffer," it involves planting some kind of vegetation along the edge of a watery are, like a lake or stream, to protect the water quality.

27. **Ruminant:** An animal, such as the Belted Galloway cattle, that has a specialized stomach with four compartments and chews cud as part of the digestion process.

28. **Steer:** Castrated bull

29. **Vaccination:** The administration of a vaccine to Belted Galloway cattle to stimulate their immune system and protect them against specific diseases.

30. **Weaning:** The process of separating a calf from its mother and transitioning it to a diet of solid food, typically occurring at around six to eight months of age in Belted Galloway cattle.

References

Beef Quality Assurance (BQA). (n.d.). *What is BQA?* https://www.bqa.org/

Beef-Cattle. (2019, September 3). *Livestock biosecurity*. Beef-cattle.extension.org. https://beef-cattle.extension.org/livestock-biosecurity/

Belted Galloway. (2022, September 29). The Cattle Site. https://www.thecattlesite.com/breeds/beef/9/belted-galloway

The Belted Galloway Cattle Society. (n.d.). *History*. Belted Galloway. https://www.beltedgalloways.co.uk/history/

Belted Galloway Society. (n.d.). Home page. https://beltie.org/

CattleLink vs. CattleMax vs. Cow Sense Suite comparison chart. (n.d.). SourceForge. Retrieved May 22, 2024 https://sourceforge.net/software/compare/CattleLink-vs-CattleMax-vs-Cow-Sense-Suite/

Cockram, M. S. (2019). Fitness of animals for transport to slaughter. *The Canadian Veterinary Journal, 60*(4), 423–429. https://www.ncbi.nlm.nih.gov/pmc/articles/PMC6417610/

Darris, D. (2003). *Considerations for establishing native grasses from seed for restoration, revegetation, and erosion control in western Washington and western Oregon*. National Resources Conservation Service (NRCS); United States Department of Agriculture (USDA). https://www.nrcs.usda.gov/plantmaterials/orpmctn5165.pdf

Gunnars, K. (2019, December 4). *Grass-fed vs. grain-fed beef—What's the difference?* Healthline. https://www.healthline.com/nutrition/grass-fed-vs-grain-fed-beef

Homer, M. (2024, March 18). *Rodeo by the numbers: 2024 HLSR shatters attendance, auction records*. Khou-TV. https://www.khou.com/article/entertainment/events/rodeohouston/houston-rodeo-new-record-attendance-auctions/285-157198ae-6122-4ca7-a72d-bc0a4e437d33

Intelligence Hive. (2024, April 22). *Premium beef market: Trends, forecast, and competitive analysis to 2031*. LinkedIn. https://www.linkedin.com/pulse/premium-beef-market-trends-forecast-competitive-analysis-xp5jc/

The Livestock Conservancy. (n.d.). *Belted Galloway cattle*. https://livestockconservancy.org/heritage-breeds/heritage-breeds-list/belted-galloway-cattle/

Michigan State University Extension. (2024, April 12). *Biosecurity for dairy and beef cattle farms*. https://www.canr.msu.edu/resources/biosecurity-for-dairy-and-beef-cattle-farms

Oltenacu, P., & Broom, D. M. (2010). The impact of genetic selection for increased milk yield on the welfare of dairy cows. *Animal Welfare, 19*(S1) 39–49. https://doi.org/10.1017/S0962728600002220

Rebel Ranch. (n.d.). *Rebel Ranch: Cattle ranch & honey bee apiary*. https://rebelranch.net

6 tips for properly vaccinating beef cattle. (2021, February 5). Farm Progress. https://www.farmprogress.com/management/6-tips-for-properly-vaccinating-beef-cattle

Tips for cooking your grassfed beef. (2024). Rebel Ranch. https://rebelranch.net/pdf-files/cooking_grass-fed_beef.pdf

Image References

All photos provided by Twila Weiss and JD Weiss, authors, photographers, and co-owners of Rebel Ranch.

Meet JD and Twila Weiss

(This is our cattle brand, RE Bell *[pronounced "rebel"]. It appears reversed here, but that's how it's branded on the animal's hide for proper reading.)*

JD and Twila Weiss are the hardworking co-owners of Rebel Ranch Publishing and the Rebel Ranch. These talented individuals operate a Belted Galloway cattle ranch and Honeybee Apiary. Together, JD and Twila have achieved recognition not only in the United States but also on an international scale. Their diverse collection of books can be easily obtained from online platforms or local physical bookstores. With years of writing under their belt, they bring a unique perspective to the non-fiction, fiction, and children's book genres. Moreover, their skillset extends beyond traditional writing, encompassing the creation of official documents and innovative papers that set them apart in their field.

Writing Expertise

One cannot help but be impressed by the depth of experience exhibited by JD and Twila Weiss. Their journey as published authors has given them valuable insights into the intricacies of the written word. From shaping compelling narratives in fiction to elucidating complex concepts in non-fiction, the dynamic duo does it all with finesse. Their ability to tackle various writing styles and genres reflects their adaptability as writers.

Ongoing Projects

JD and Twila are fully immersed in the creative process, diligently working on multiple book manuscripts. This stage of their journey highlights their dedication to their craft and their unyielding commitment to producing high-quality content consistently. It is this passion for storytelling that drives them forward, pushing the boundaries of their creativity to deliver captivating narratives that resonate with readers across the globe.